Springer Proceedings in Business and Economics

Springer Proceedings in Business and Economics brings the most current research presented at conferences and workshops to a global readership. The series features volumes (in electronic and print formats) of selected contributions from conferences in all areas of economics, business, management, and finance. In addition to an overall evaluation by the publisher of the topical interest, scientific quality, and timeliness of each volume, each contribution is refereed to standards comparable to those of leading journals, resulting in authoritative contributions to the respective fields. Springer's production and distribution infrastructure ensures rapid publication and wide circulation of the latest developments in the most compelling and promising areas of research today.

The editorial development of volumes may be managed using Springer Nature's innovative EquinOCS, a proven online conference proceedings submission, management and review system. This system is designed to ensure an efficient timeline for your publication, making Springer Proceedings in Business and Economics the premier series to publish your workshop or conference volume.

This book series is indexed in SCOPUS.

Adina Letiția Negrușa • Monica Maria Coroş
Editors

Innovation, Sustainability, and Growth in a VUCA Environment

Proceedings of the 5th International Conference on Modern Trends in Business, Hospitality and Tourism (ICMTBHT), Cluj-Napoca, Romania, May 22-24, 2025

Springer

Editors
Adina Letiţia Negruşa
Faculty of Business
Babeş-Bolyai University
Cluj-Napoca, Romania

Monica Maria Coroş
Faculty of Business
Babeş-Bolyai University
Cluj-Napoca, Romania

ISSN 2198-7246 ISSN 2198-7254 (electronic)
Springer Proceedings in Business and Economics
ISBN 978-3-032-11638-3 ISBN 978-3-032-11639-0 (eBook)
https://doi.org/10.1007/978-3-032-11639-0

This work was supported by BABES-BOLYAI UNIVERSITY

Volume financed from the Special Situations' Fund of the Ministry of Education and Research of Romania/Volum finanţat din Fondul de Situaţii Speciale al Ministerului Educaţiei şi Cercetării al României

International Conference on Modern Trends in Business Hospitality and Tourism ictbht ictbht2025 ICMTBHT 5 Napoca Romania 2025 5 22 2025 5 24 https://tbs.ubbcluj.ro/conference/index.html

This Springer imprint is published by the registered company Springer Nature Switzerland AG
The registered company address is: Gewerbestrasse 11, 6330 Cham, Switzerland

Preface

It is with great pleasure that I write the preface for this edited volume featuring 15 contributions from the 2025 International Conference on Modern Trends in Business, Hospitality and Tourism hosted by the Faculty of Business, Babeş-Bolyai University in Cluj-Napoca, Romania. This volume represents the fifth edition of the conference and the fourth time the proceedings have been published as part of the Springer Proceedings in Business and Economics.

The conference theme *of Innovation, Sustainability, and Growth in a VUCA Environment* seems particularly relevant in the world today. As someone more familiar with the related phrase of "wicked problems," I am a little embarrassed to admit I had to look up VUCA—volatility, uncertainty, complexity, and ambiguity—when I was invited to be one of the Keynotes for the conference. What quickly became very apparent at the conference was how much VUCA resonated across the many insightful, thought-provoking, and interesting papers and presentations.

For me, one of the strengths of the conference and these Proceedings is the interdisciplinary engagement with the conference theme and sub-themes. The 91 papers presented at the conference, including the 15 papers in this volume, represent 138 researchers and practitioners from 31 countries and 53 universities and cover areas of economics, management, marketing, and tourism research.

The conference featured seven keynote presentations across tourism, economics, and human resources management. Three of the keynotes have contributions included in this volume. The first keynote at the conference, Professor Greg Richards from Tilburg University in the Netherlands, presents a paper in this volume that looks at emerging business models in rural areas through curated cultural tourism experiences. Exploring how modes of curation are altering the development of rural areas, the paper concludes that the staging of experiences, co-creation of culture, and growth in digitalization are significantly altering rural areas to accommodate the tastes of urban tourists. The second keynote contribution comes from Professor Sébastien Point from Université de Strasbourg in France. Along with Elisabeta Butoi from the host university of Babeş-Bolyai University, Romania, Sébastien presents a paper in this volume focused on workplace well-being, innovation, and embodiment. The paper challenges existing approaches to well-being in the workplaces and instead suggests a more integrated framework that emphasizes embodied, relational, and technological dimensions. The third and final keynote contribution in this volume is from Andrei Rădulescu, PhD, from the Institute for World Economy, Romanian Academy. His paper focuses on examining the macroeconomic climate from the perspective of the global and the local. His analysis suggests that current global events and shocks (such as geopolitical tensions and the COVID pandemic) have impacted the growth rate of the world economy. His findings reveal that the likelihood of a global economic crisis in the near future has significantly increased and estimations of potential growth for specific economies, including the USA and Romania, are weaker than pre-COVID performances.

The four other keynotes were Professor Benno Torgler (Queensland University of Technology, Australia) who talked about the opportunities and challenges in behavioral economics; Professor Stanislav Ivanov (Varna University of Management, Bulgaria) who told the conference about the economics of service robots in tourism and hospitality; Professor Morris Altman (University of Dundee, UK) who focused on class and sustainability modeling when thinking about implications for equity and well-being; and myself (Tara Duncan, Thompson Rivers University, Canada) where I considered the future of decency and dignity in tourism and hospitality work.

Along with the seven keynotes, the conference was packed with the 91 papers spread over concurrent sessions and all with hybrid (Teams) links. Papers at the conference included presentations from over 30 students and I feel fortunate to have attended a number of the student sessions. From those 91 papers come the remaining 12 papers in this volume. The topics range from the economic impact of US-China tensions (Silvo Dajčman and Dejan Romih, University of Maribor, Slovenia; Valentin Toader, Babeş-Bolyai University, Romania) to leadership structures and partial state ownership in the hospitality sector (Flávio Ferreira, Polytechnic of Porto, Portugal, and Jorge Umbelino, Estoril Higher Institute for Tourism & Hotel Studies, Portugal) to smart technologies and data sharing in tourism businesses (Tomáš Gajdošík and Beáta Vaneková, Matej Bel University in Banská Bystrica, Slovakia). The papers in this volume offer interesting, informative, and timely contributions to consider the conference theme of innovation, sustainability, and growth in a VUCA environment. This volume can be a useful resource for researchers, practitioners, and postgraduate students alike, and as the authors in this volume come from different academic backgrounds and take diverse approaches to the conference theme, it offers a unique multi- and interdisciplinary perspective to research in this area.

Many thanks to the Scientific Committee and reviewers who took time to review abstracts and paper for the conference and the Proceedings. Many thanks also go to the conference organizing team at the Faculty of Business, Babeş-Bolyai University. The team did an amazing job of carefully organizing a busy few days of presentations, networking, social events, and more. It is obvious from the previous Proceedings and this volume that this conference will continue to grow. Echoing the preface from the 2024 Proceedings, the conference successfully achieved its goals of bringing together a diverse range of researchers, practitioners, and students to exchange ideas, learn more about ongoing research, and discuss recent and future trends in relations to the challenges and solutions facing the fields of business, hospitality, and tourism.

I would like to conclude with a few personal reflections about the conference. Firstly, special thanks go to Monica Maria Coroş who invited me to be a Keynote at this conference and hosted my visit. It was an honor to be about to solidify connections and build friendship(s) through shared—and new—research interests. Secondly, the interdisciplinary nature of the conference was refreshing and motivating. Through several of the presentations and especially Benno Torgler's and Morris Altman's keynotes, I have realized I now need to explore the field of economics in more detail. Thank you for the inspiration—without the variety of disciplines represented at this conference, I would not have found new avenues to explore. Finally, I look forward to future iterations of the conference as I know this conference will go from strength to strength and I hope many of my colleagues from around the world will discover it. The academic content was stimulating and the welcome of our Romanian hosts and their hospitality throughout the conference and post-conference trip (which was amazing) were beyond compare. Thank you very much to everyone involved.

Thompson Rovers University, Kamloops, BC, Canada Tara Duncan
taduncan@tru.ca

Contents

Chapter 1
Macroeconomic Climate: From World Economy to National Economy

Andrei Rădulescu

Abstract The exogenous shocks (the coronavirus pandemic and the geopolitical tensions) and their consequences have had an important impact on the macroeconomic climate in recent years. The annual growth pace of the world economy decelerated to only 3.3% in 2024, the weakest dynamics since the pandemic year 2020, according to the estimates of the International Monetary Fund (IMF, *World economic outlook—A critical juncture amid policy shifts*, 2025a). Furthermore, in the first months of 2025, the uncertainty increased to record high levels due to the major shifts in trade policy in the USA, the largest economy in the world. In this context, the probability of the outbreak of a new global economic crisis in the short run has significantly increased, being estimated at around 30% by the experts of the IMF in April 2025. In this paper, we analyze the recent macroeconomic developments from a perspective of the world economy, the Euroland economy, and the Romanian economy. Furthermore, we implement standard econometric tools (the Hodrick-Prescott filter) and use the IMF database (annual statistics for the period 2000–2025) in order to estimate the annual potential growth rate for the world economy, the USA, Euroland, and Romania. According to the results of our estimates, the annual potential growth rate for world economic activity in 2025 is 3.1%, below the average for the period 2000–2019 (3.6%). There is a gap between the USA and Euroland in terms of the annual potential growth rate of economic activity, with estimates for 2025 at 2.3% and 1.3%, respectively. This high level of the gap is explained by the fact that the economy of Euroland paid a higher invoice since the outbreak of the coronavirus pandemic. In the case of Romania, the annual potential growth is estimated at only 2.7% in 2025, the weakest performance since 2000, as the country is confronted with high levels of the twin deficits.

Keywords World economy · Potential output · Romanian economy

A. Rădulescu (✉)
Institute for World Economy, Romanian Academy, Bucharest, Romania

A. L. Negrusa, M. M. Coroş (eds.), *Innovation, Sustainability, and Growth in a VUCA Environment*, Springer Proceedings in Business and Economics,
https://doi.org/10.1007/978-3-032-11639-0_1

1.1 Introduction

The macroeconomic climate has recently deteriorated, given the consequences of the exogenous shocks (the coronavirus pandemic and the geopolitical tensions), including the geoeconomic fragmentation and the upward trend for the real financing costs.

According to the 2025 Spring edition of the World Economic Outlook (WEO) of the International Monetary Fund (IMF, 2025a) the world economy (world GDP, constant prices) grew by an annual pace of 3.3% in 2024, slowing down from 3.5% in 2023. This is the weakest performance since the pandemic year 2020, as reflected in the following chart.

The evolution was determined by the deterioration of the annual growth pace in the emerging and developing economies (EMDEs), from 4.7% in 2023 to 4.3% in 2024, the slowest since 2022, as the investment climate has been affected by the upward trend for the real financing costs (nominal costs adjusted for inflation).

On the other hand, the annual growth pace of the advanced economies (AE) accelerated from 1.7% in 2023 to 1.8% in 2024, the strongest since 2022, due to the improvement of the activity in Euroland—annual growth pace of 0.9% in 2024, up from 0.4% in 2023.

We point out the positive momentum in the US economy, with an annual growth pace of 2.8% in 2024, very close to that in 2023 (2.9%), thus confirming the continuity of the ***American Exceptionalism*** phenomenon, recently described by experts (Carson, 2025) (Fig. 1.1).

The deteriorating trend of world economic activity continued in the first months of 2025, an evolution strongly influenced by the increase in the degree of uncertainty, determined by the major shifts in trade policy (Szlezak et al., 2025).

In fact, the degree of economic policy uncertainty in the USA climbed to record high levels in April 2025 (outpacing the level at the moment of the outbreak of the

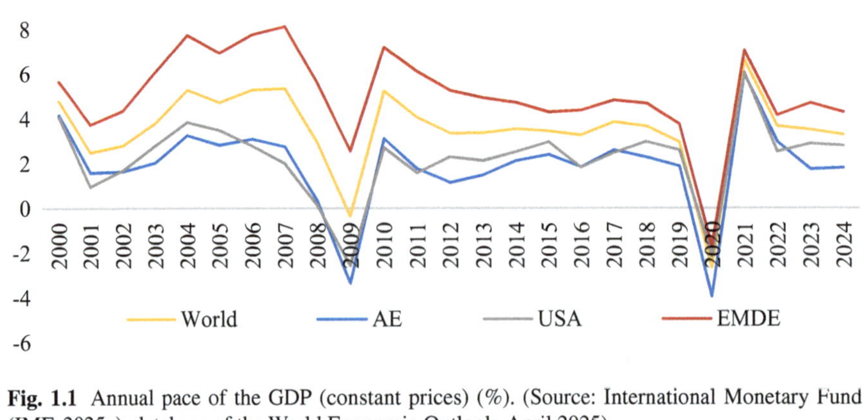

Fig. 1.1 Annual pace of the GDP (constant prices) (%). (Source: International Monetary Fund (IMF, 2025a), database of the World Economic Outlook, April 2025)

coronavirus pandemic), after the White House (2025) announced the increase in the tariffs on imports of goods at the beginning of the month, according to the statistics of the Federal Reserve Bank of Saint Louis (2025), as can be noticed in the following chart. If the announced measures were implemented, the annual average trade tariff in the USA would increase to the highest level since the beginning of the last century (Fig. 1.2).

In this context, the probability of the outbreak of a recession in the world economy in the short run (next 12 months) has recently increased, being estimated at 30% by the experts of the International Monetary Fund (IMF, 2025b).

The rest of the paper has the following structure: the next chapter briefly presents the literature review; the methodology is described in the third chapter; the results are interpreted in the fourth chapter; and the conclusions are drawn in the last chapter.

1.2 Literature Review

The deterioration of the world macroeconomic climate in recent years was determined by several factors, including the consequences of the exogenous shocks, the persistence of geopolitical tensions at a high level, climate change, and the unfavorable prospects for international trade.

For instance, after the outbreak of the crisis in Ukraine a fragmentation in international trade and investment flows can be noticed, according to the research developed by several experts (Gopinath et al., 2024). Employing a gravity model and dividing the countries into three blocks (US-leaning, China-leaning, and the non-aligned group) the research estimated that after the outbreak of the crisis in Ukraine until the end of 2023, the flows of trade and foreign direct investments between

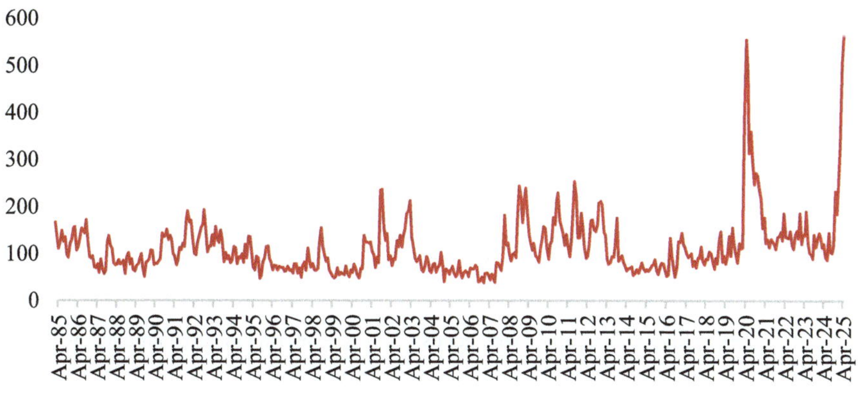

Fig. 1.2 Economic Policy Uncertainty Index in the USA. (Source: Federal Reserve Saint Louis database, 2025)

these blocks contracted by 12%, and 20% respectively, more than the flows within the blocks.

In fact, the volume of international trade in goods grew by an annual pace of only 3.1% in 2022 (slowing down from 10.0% in 2021), contracted by an annual rate of 0.8% in 2023, and increased by an annual dynamic of only 2.5% in 2024, according to the statistics of the Netherlands Bureau for Economic Policy Analysis (CPB, 2025).

The adjustments of the international trade in goods since the outbreak of the crisis in Ukraine have had a negative impact on the world industry. The volume of the global industrial production grew by annual paces of 3.0% in 2022, 1.0% in 2023, and 1.7% in 2024, according to the CPB statistics, as can be noticed in the following chart (Fig. 1.3).

In this context, the Managing Director of the International Monetary Fund (IMF, 2024) underlined that international trade ceased to be a ***powerful engine of growth***.

On the other hand, the rising public debt also contributed to the weak performance of the world economy recently. Several papers documented the fact that, in general, the increase in the public debt has a negative impact on the growth pace of the economic activity (San & Chin, 2023; de Soyres et al., 2022).

We point out that the budget deficit/GDP ratio in the world widened from 3.7% in 2022 to 4.9% in 2023 and 5.0% in 2024, the highest level since the pandemic year 2021, according to the database of the International Monetary Fund (IMF, 2025a). This evolution was mainly determined by the increase of the budget deficit/GDP ratio in the advanced economies, from 2.9% in 2022 (the year of the outbreak of the crisis in Ukraine) to 4.6% in 2023, and 4.7% in 2024. In the emerging and developing economies, the budget deficit/GDP ratio intensified from 2.3% in 2022 to 2.5% in 2023, and 2.6% in 2024.

In this context, the public debt/GDP ratio increased from 89.9% in 2022 to 91.3% in 2023 and 92.3% in 2024 at the world level, the highest since 2021. In the

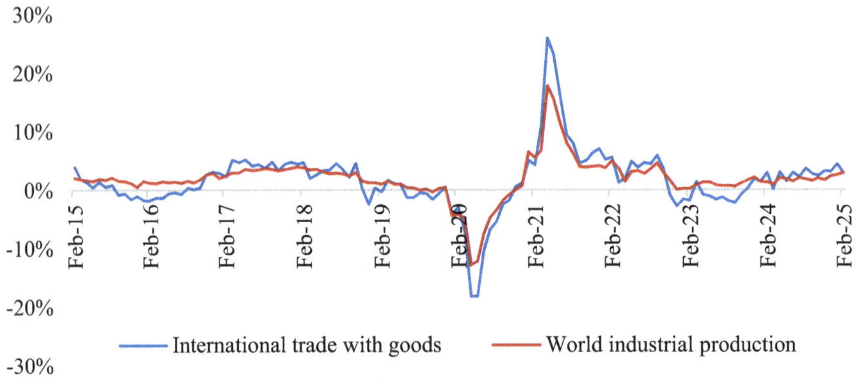

Fig. 1.3 International trade with goods vs. world industrial production (annual rate). (Source: Netherlands Bureau for Economic Policy Analysis (CPB), 2025)

developed economies, the public debt/GDP ratio presented levels of 109.3% in 2022, 108.2% in 2023, and 108.5% in 2024. On the other hand, in the emerging and developing economies, this indicator climbed from 63.2% in 2021 to 63.4% in 2022, 67.4% in 2023, and 69.5% in 2024, respectively.

The deterioration of public finance at the world level after the outbreak of the crisis in Ukraine was determined by several factors, including the increase in the military spending, the implementation of programs to counter the increase in energy prices, and the upward trend in the real financing costs. For instance, the military spending in the world rose from USD 2.1tn in 2022 to USD 2.4tn in 2023, and USD 2.7tn in 2024, according to the database of the Stockholm International Peace Research Institute (SIPRI, 2025).

In a recent paper, Ginn and Saadaoui (2025) implemented an augmented Taylor rule on a panel of 18 economies and found that interest rates increase in the mid-run after the outbreak of a geopolitical risk shock. In fact, the real interest rate on 10YR maturity in the USA (a benchmark for the long-run financing costs in the largest economy in the world, but also for the global economy) rose from negative territory in 2020 and 2021 to 1.82% in 2024, the highest level since 2007, according to the database of the Federal Reserve Saint Louis (FED, 2025), as can be noticed in the following chart (Fig. 1.4).

Last but not least, the world economic activity has recently continued to be negatively affected by climate change, as 2024 was the warmest year, with 1.55 °C above the pre-industrial level, according to the estimates of the World Meteorological Organisation (WMO, 2025).

For instance, the increase of the global temperature by 1 °C would determine the intensification of the extreme climatic shocks, leading to a permanent decline of the world GDP of up to 12% at the peak, similar to what happened during the Great Depression (Bilal & Känzig, 2024). Furthermore, the world would be confronted with a decline in income by 19% until 2050 irrespective of the future emission choices (Kotz et al., 2024).

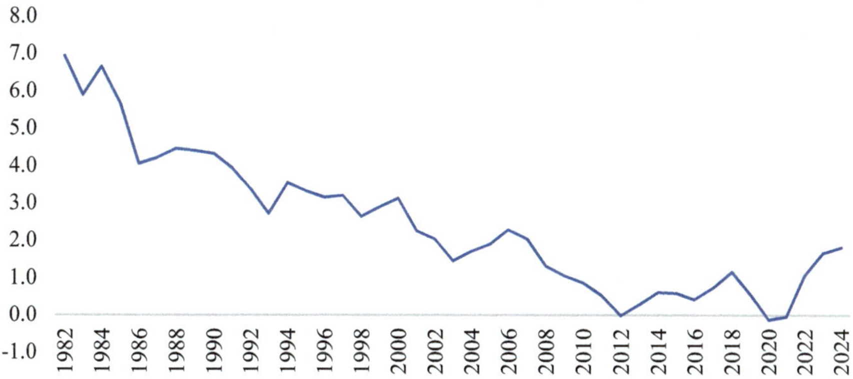

Fig. 1.4 10 YR real interest rate in the USA (%). (Source: Federal Reserve Saint Louis, 2025)

1.3 Research Methodology

In this paper we implement standard econometric tools and use the database of the International Monetary Fund (IMF, 2025a) in order to estimate the annual rate of potential output for the world economy (distinguishing between the advanced and emerging and developing countries), the USA, the Euroland, and Romania during the period 2000–2025.

We opted for the Hodrick-Prescott filter, a transparent, widely used, and simple method to distinguish between the structural and the cyclical components of the annual rate of economic activity (Bouthevillain et al., 2001).

Any macroeconomic variable can be expressed as the sum of the structural component (the trend, which is determined by structural elements) and a cyclical component (the cycle, which is determined by cyclical factors).

In their seminal paper, Hodrick and Prescott (1997) estimated the structural component of the GDP by using the following relation:

$$\text{Min} \sum_{t=1}^{T} \left(\ln Y_t - \ln Y_t^* \right)^2 + \lambda \sum_{t=2}^{T-1} \left(\left(\ln Y_{t+1}^* - \ln Y_t^* \right) - \left(\ln Y_t^* - \ln Y_{t-1}^* \right) \right)^2 \quad (1.1)$$

In this relation Y_t, $Y_{t,}^*$ and λ are the GDP, its trend, and a smoothness parameter, the lower the value of this, the closer the output to its potential. In fact, if the value of this smoothness parameter is 0 the GDP is equal to its trend component.

We also point out the shortcomings of this econometric method, the so-called leakage effects and compression effects, as it is extremely difficult to perfectly distinguish between the structural and the cyclical components of the macroeconomic variables.

In our paper, we used a value of 100 for this smoothness parameter, as recommended by Hodrick & Prescott while working with annual data.

We used the database of the International Monetary Fund (IMF, 2025a), selecting the annual rate of the GDP for the world economy, advanced economies, emerging and developing countries, the USA, the Euroland, and Romania.

On the one hand, we used the estimates of the international institutions for the period 2000–2024.

On the other hand, we considered the forecasts of this institution for the interval 2025–2030, updated in April 2025.

In this paper, the econometric software E-Views was used.

1.4 Results and Discussions

According to the results of the econometric estimates, the annual pace of the potential output in the world economy is around 3.1% in 2025, as can be noticed in the following chart. It is close to the lowest level since the pandemic year 2020 (which was the minimum for the period under analysis), and below the average for the period 2000–2019 (3.6%).

Furthermore, we point out the downward trend for this indicator since the outbreak of the Great Financial Crisis, an unprecedented systemic shock, with a symmetric impact, and asymmetric consequences (Fig. 1.5).

In other words, the shocks over the past years have had an impact on the evolution of the potential output in the world economy, despite the massive mobilization of resources by policymakers in order to counter the consequences of these.

On the other hand, we point out the divergence between the advanced economies and the emerging and developing countries in the evolution of the annual pace of the potential output, according to the results of our econometric analysis, reflected in the following chart.

The annual pace of the potential output in the former group of countries decelerated from 1.84% in 2024 to 1.80% in 2025, the lowest level since 2019, the year before the outbreak of the coronavirus pandemic, which was the worst health crisis in the world in more than a century.

It is a level slightly below the annual average for the period 2000–2019 (1.9%), but higher than the level in 2010 (1.4%, the minimum for the past decades), when Euroland (which contributes by around 85% to the GDP of the European Union) started to be confronted by the second wave of the Great Financial Crisis, the public debt crisis.

The deterioration of the annual rate of the potential output across the developed countries in recent years was also determined by the unfavorable developments in the Euroland, as the economy of the region was the most affected by the consequences of the crisis in Ukraine.

The results of our econometric analysis point to the slowing down of the annual rate of potential output in the Euroland, from 1.33% in 2024 to only 1.30% in 2025, the minimum since the pandemic year 2020, as reflected in the following chart. The lowest level for this indicator during the period 2000–2025 was hit in 2011 (0.7%),

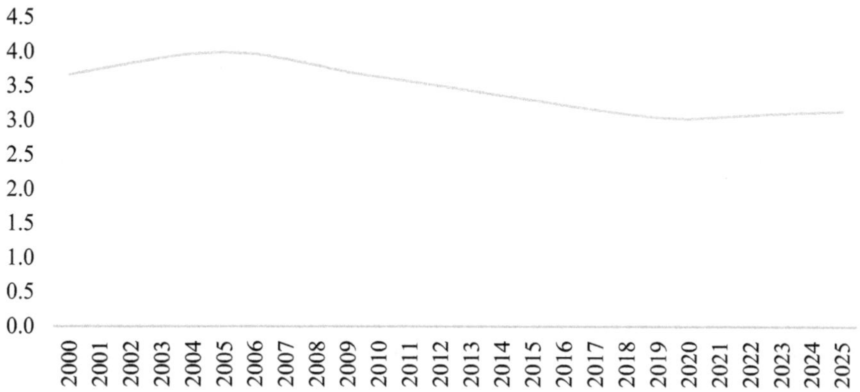

Fig. 1.5 The annual rate of potential output, world economy (%). (Source: Results of the econometric analysis, using the IMF database, 2025)

when the region was confronted with the public debt crisis, the second wave of the Great Financial Crisis. In fact, the annual rate of the potential output in the region had significantly deteriorated since the beginning of the 2000s, when it was estimated at 2.3%, as all the global shocks that occurred over the past decades have revealed the structural weakness of the European monetary integration process.

Furthermore, the current low level of the annual pace of the potential output in the Euroland is very close to the annual average recorded for the interval 2000–2019, as the economy of the region was severely affected by the recent unprecedented global shocks (the coronavirus pandemic and the geopolitical tensions) and their consequences since 2020.

In this respect, we point out that the economy of Euroland continued to be confronted in recent years with the long-lasting challenges, the deterioration of international competitiveness, and the discontinuity of economic convergence. This was also emphasized by several economists, who advanced solutions for the transformation of Europe (Rădulescu, 2024a). The Romanian economist proposes annual targets for the most important structural indicators, starting with labor productivity, continuing with digitalization, R&D allocation, development of the financial markets, transition to green economy, reduction of the regulatory burden, acceleration of the green economy, improvement of the economic diplomacy, and ending with closing the gaps among the countries in terms of competitiveness and the improvement of the communication among the stakeholders.

We emphasize the fact that the economic growth model in Euroland over the past decades was dependent on the imports of cheap oil and gas from Russia and on the exports of high-value-added goods to China. These elements ceased to be the engines of economic growth in the region.

On the other hand, in the USA (the largest economy in the world, with a nominal GDP close to USD 30tn in 1Q 2025, according to the estimates of the Bureau of Economic Analysis, 2025) the annual pace of the potential output continues to be higher than in the Euroland in 2025. The gap in terms of the annual rate of potential GDP between the USA and the Euroland is estimated at over 1 pp for the 14th year in a row in 2025, according to the results of our econometric analysis, as reflected in the following chart.

The annual pace of the potential output in the USA is estimated at 2.3% for 2025. This is the lowest level since 2018, but it is above the average of the period 2000–2019 (estimated at 2.2%), as the largest economy in the world (contributing more than 26% to the global GDP) implemented an unprecedented fiscal impulse since the outbreak of the coronavirus pandemic (Fig. 1.6).

The developments in terms of the annual pace of the potential output (the high level of the gap between the USA and the Euroland) determined a divergent path for the weight of the nominal GDP in the world economy. On the one hand, the share of the USA economy in the global GDP consolidated above 25% in recent years. On the other hand, the share of the Euroland GDP in the global GDP declined to below 15% at present.

In this respect, we point out that unless structural reforms are accelerated in the Euroland, its share in the world economy may decline to below 10% by the end of

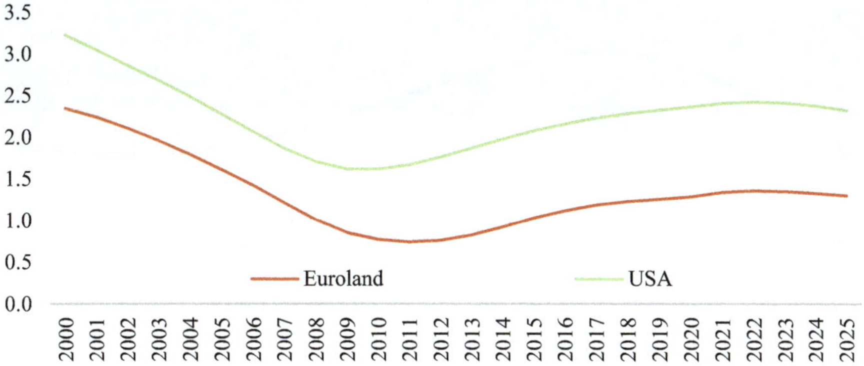

Fig. 1.6 The annual rate of potential output—the USA vs. the Euroland (%). (Source: Results of the econometric analysis, using the IMF database, 2025)

this decade. We must underline here that since the new European Commission started its current mandate, some important reforms have been launched to support the performance of the economy of the European Union, including the Competitiveness Compass (focused on reducing the regulatory burden) and the improvement of the security of the region (including ReArm EU) (European Commission, 2025a, b)

In the emerging and developing countries, the annual pace of the potential output improved from 3.94% in 2024 to 3.96% in 2025. This is the best performance since 2019, the year before the coronavirus pandemic, as can be noticed in the Fig. 1.7.

However, at present, the annual rate of the potential GDP in this group of countries is significantly below the annual average for the period 2000–2019, estimated at 5.3%, according to the results of our econometric analysis.

As regards the Romanian economy (the second largest in Central and Eastern Europe and the 12th largest in the European Union, with a nominal dimension of over EUR 350bn in 2024) (Eurostat, 2025) the results of our econometric estimates point to the deceleration of the annual rate of the potential output to 2.7% in 2025. This is the lowest level since 2000, as reflected in the following chart. Furthermore, the current estimated level of the annual pace of the potential GDP in Romania is below the annual average for the period 2000–2019 (3.5%) (Fig. 1.8).

The deterioration of the annual rate of the potential output in Romania in recent years was determined by several factors, including the discontinuity of structural reforms, the widening of twin deficits, and the intensification of risk perception in the context of unprecedented political tensions (end of 2024 and the first half of 2025).

According to the estimates (Eurostat, 2025) the Romanian GDP climbed for the fourth year in a row in 2024, by an annual rate of 0.8%, the weakest pace since 2020.

This evolution was determined by the contraction of gross fixed capital formation at the most severe pace since 2013 (annual rate of 3.3%), in the context of the intensifying political tensions, the widening of the budget deficit, and the

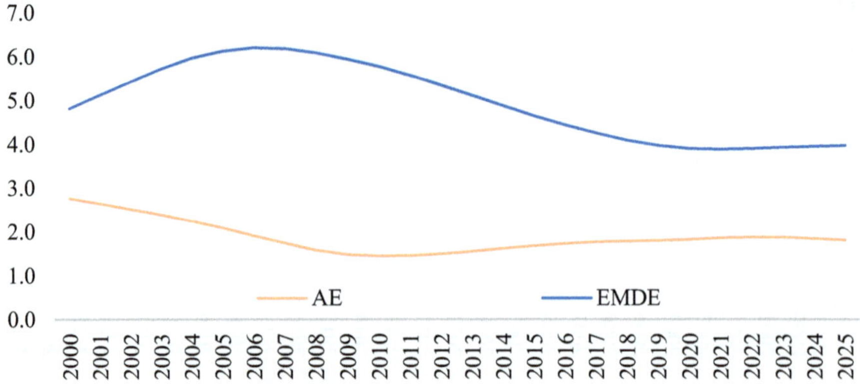

Fig. 1.7 The annual rate of potential output—advanced economies vs. emerging and developing economies (%). (Source: Results of the econometric analysis, using the IMF database, 2025)

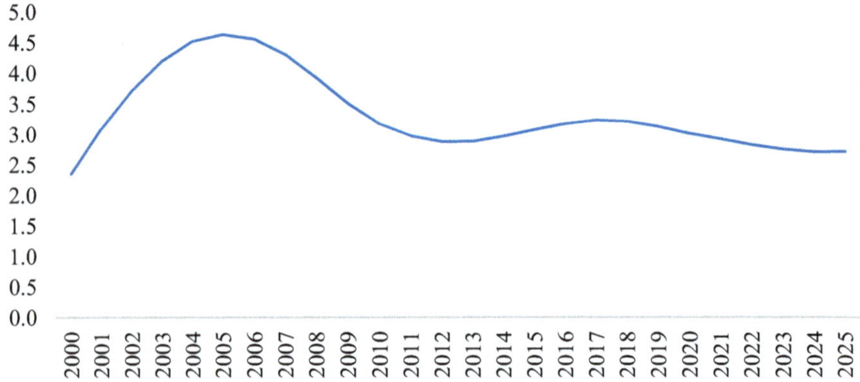

Fig. 1.8 The annual rate of potential output in Romania (%). (Source: Results of the econometric analysis, using the IMF database, 2025)

deterioration of international competitiveness. The engine of the economy adjusted by an annual rate of 16.1% in Q4 2024, the most severe contraction since Q4 2009.

On the other hand, the annual pace of private consumption (the main component of the GDP) accelerated to 6.0% in 2024, the strongest since 2021, on the back of the expansionary income policy.

The public consumption rose for the second year in a row in 2024, but the annual pace decelerated to 0.7%.

The contribution of net foreign demand to the annual pace of the economy was negative last year, as exports contracted for the second year in a row, by an intensifying annual rate of 3.1%, while imports rebounded, climbing by an annual rate of 3.8%.

From the supply-side perspective, value added in the domestic economy grew by an annual rate of only 0.1% in 2024, slowing down from 2.1% in 2023, an evolution diverging from the EU developments (the annual pace accelerated from 0.7% in 2023 to 1.0% in 2024).

There can be noticed the contraction of value added in construction and agriculture by annual paces of 2.6% and 5.9%, respectively, in 2024.

The value added also decreased last year in industry, financial and insurance activities, and real estate activities, by annual paces of 0.2%, 0.3%, and 1.3%, respectively.

On the other hand, the value added grew in IT&C by an annual rate of 0.3% and in wholesale and retail trade, transport, accommodation, and food service activities by an annual rate of 1.7% in 2024.

However, the country continued the European economic convergence process, the GDP/capita at purchasing power standards representing 78.3% of the EU average in 2024. This level is higher by around 50 percentage points compared to the level in the mid-1990s, Romania being the champion of the real economic convergence process.

On the other hand, we point out that Romania needs a new model of growth and development in the future, as the model determined and supported by the forces of the EU integration process has reached its limits.

A new model of growth and development should be focused on the incorporation of technological progress, digitalization, and reform of the public sector, development of capital markets, consolidation of public finances in a sustainable way, and significantly improving the allocation for research and development.

1.5 Conclusions

At the beginning of 2025, the world economy continues to be confronted with the consequences of the previous shocks, namely the coronavirus pandemic and the intensification of the geopolitical tensions following the outbreak of the crisis in Ukraine. Furthermore, a new shock hit the world economy in the context of the changes in terms of the trade policy in the USA.

In this context, the growth pace of the world economy significantly decelerated in the month of April 2025, recording the lowest level since the end of 2023, according to the PMI Composite indicator (S&P Global, 2025). This indicator, published on a monthly basis, reflects the developments in new orders, production, employment, inventories, prices, and short-term outlook in both manufacturing and services, with the 50-point threshold separating expansion from contraction.

Therefore, the annual potential growth rate of the world economy is expected to deteriorate, while the divergences among the largest economic blocs in the world are forecasted to widen in 2025, confirming the developments in recent years, as reflected by the results of our analysis. In other words, the world economy would diverge from the pre-pandemic period as regards the annual rate of potential output in 2025.

In this respect, we point out that it is more important than ever in the post-Second World War period to have a new Bretton Woods momentum—negotiations among the representatives of the largest economies for a sustainable New World Order, based on rules, promotion of peace, trade, economic diplomacy, and win-win situations (Rădulescu, 2024b).

The results of the current analysis present a particular importance for the national economy (Romania), where the macroeconomic climate has deteriorated over the past few quarters, as the difficult international environment and internal tensions have had a negative impact on companies' investment plans.

The annual growth rate of economic activity slowed to 0.8% in 2024 (the weakest performance since the pandemic year 2020) and to 0.2% in the first quarter of 2025, according to preliminary estimates (National Institute of Statistics, 2025).

At the same time, between 2019 and 2024, the domestic economy grew at an average annual rate of only 1.8%—the weakest pace since the last decade of the twentieth century, during the Caritas-style economy and the difficult and consequential transition toward a market economy.

The recent deterioration in the pace of economic activity, combined with the intensification of twin deficits and the slowdown in the potential annual growth rate (toward the lowest level since 2000, according to our econometric estimates), are factors indicating that Romania needs a new model for growth and development.

One that is different from the model of the past decades, which was dominated by the forces of integration into European and global economic circuits.

A new model of growth and development must prioritize measures for the proper allocation of resources (which the Romanian economy possesses), with an emphasis on continuous structural reforms and the targeting of ambitious but achievable objectives, such as creating a competitive regional, European, and global hub for international trade and investment flows, energy, the tech sector, and high-value-added products and services.

Romania has the resources (human, financial, digital, agricultural), but what is needed is prioritization of measures and significant changes in terms of governance, implementation, and accountability, so that in the next decade we can speak of an economy in the top half of the European Union in terms of competitiveness, adoption of the Digital and Artificial Intelligence Revolutions, research and development allocations, share of the population with higher education, development level of the financial market, regulatory quality, etc.

At the end of this paper, we point out the limitations of the study. On the one hand, we point out the revisions of the estimates regarding the evolution of the GDP, especially in the case of Romania. On the other hand, we emphasize the drawbacks of the econometric filter implemented in our analysis. In this context, an alternative method would be the implementation of the Cobb-Douglas production function in order to estimate the contribution of the production factors to the annual rate of potential output. Last but not least, this study does not estimate precisely the impact on the evolution of the annual pace of potential output of each of the following factors: the high level of uncertainty, the trade tensions, and the geopolitical tensions.

References

Bilal, A., & Känzig, D. R. (2024). *The macroeconomic impact of climate change: Global vs. local temperature* (NBER Working Papers, 32450). NBER. https://www.nber.org/papers/w32450

Bouthevillain, C., Cour-Thimann, P., van de Dool, G., Hernández de Cos, P., Langenus, G., Mohr, M., Momigliano, S., & Tujula, M. (2001). *Cyclically adjusted budget balances: An alternative approach* (European Central Bank Working Paper Series, 77). European Central Bank. http://www.ecb.de/pub/pdf/scpwps/ecbwp077.pdf

Bureau of Economic Analysis. (2025). *Gross domestic product, 1st Quarter 2025 (advance estimate).* Bureau of Economic Analysis. https://www.bea.gov/news/2025/gross-domestic-product-1st-quarter-2025-advance-estimate

Carson, J. (2025, March 3). Has the era of 'American exceptionalism' ended? *Haver Analytics.* https://www.haver.com/articles/has-the-era-of-american-exceptionalism-ended

de Soyres, C., Kawai, R., & Wang, M. (2022). *Public debt, & real GDP: Revisiting the impact* (IMF Working Paper, WP/22/76). IMF.

European Commission. (2025a). *A competitiveness compass for the EU* (COM 2025, 30, p. 5). https://commission.europa.eu/topics/eu-competitiveness/competitiveness-compass_en

European Commission. (2025b). *ReArm EU* (COM 2025, 0001). https://commission.europa.eu/topics/defence/future-european-defence_en

Eurostat. (2025). Gross Domestic Product (GDP) and main components (output, expenditure, and income). https://ec.europa.eu/eurostat/databrowser/view/nama_10_gdp/default/table?lang=en&category=na10.nama10.nama_10_ma

Federal Reserve Bank of Saint Louis (2025). *FRED Database.* https://fred.stlouisfed.org/

Ginn, W., & Saadaoui, J. (2025). Monetary policy reaction to geopolitical risks in unstable environments. *Macroeconomic Dynamics, 29*, 1.

Gopinath, G., Gourinchas, P. O., Presbitero, A., & Topalova, P. B. (2024). *Changing global linkages: A new cold war?* (IMF Working Paper No. 2024/076). IMF.

Hodrick, R. J., & Prescott, E. C. (1997). Postwar U.S. business cycles: An empirical investigation. *Journal of Money, Credit, and Banking, 29*(1), 1–16.

International Monetary Fund (IMF). (2024, November 16). *Difficult but Doable – Remarks by IMF Managing Director Kristalina Georgieva at the APEC Leaders' Retreat in Lima, Peru.* https://www.imf.org/en/News/Articles/2024/11/16/opening-remarks-at-apec-summit-in-peru

International Monetary Fund (IMF). (2025a). *World economic outlook – A critical juncture amid policy shifts.* IMF. https://www.imf.org/en/Publications/WEO/Issues/2025/04/22/world-economic-outlook-april-2025

International Monetary Fund (IMF). (2025b). *Press briefing transcript: Managing director's global policy agenda, spring meetings 2025.* IMF. https://www.imf.org/en/News/Articles/2025/04/24/tr-042425-managing-directors-press-briefing-on-gpa

Kotz, M., Levermann, A., & Wenz, L. (2024). The economic commitment of climate change. *Nature, 628*, 551–557. https://www.nature.com/articles/s41586-024-07219-0

Netherlands Bureau of Economic Policy Analysis (CPB) (2025). *CPB World Trade Monitor* February 2025. https://www.cpb.nl/en/cpb-world-trade-monitor-february-2025

National Institute of Statistics of Romania (NIS). (2025). *GDP evolution in the first quarter 2025 (estimated data).* NIS. https://insse.ro/cms/en/content/gdp-evolution-first-quarter-2025-estimated-data-flash

Rădulescu, A. (2024a). *Transforming Europe – A vision from the east.* https://www.brettonwoods.org/article/transforming-europe-a-vision-from-the-east

Rădulescu, A. (2024b). 80 *Years since the Bretton woods conference.* https://www.brettonwoods.org/article/80-years-since-the-bretton-woods-conference

S&P Global. (2025). *J.P. Morgan global composite PMI®.* S&P Global. https://www.pmi.spglobal.com/Public/Home/PressRelease/4d741acbec6d4b58ac75f54f11439b3b

San, C. K., & Chin, L. (2023). Impact of public debt on economic growth: A quantile regression approach. *South Asian Journal of Macroeconomics and Public Finance, 12*(2), 250–278. https://doi.org/10.1177/22779787231207218

Stockholm International Peace Research Institute (SIPRI). (2025). *SIPRI military expenditure database*. SIPRI. https://www.sipri.org/databases/milex

Szlezak, P. C., Swartz, P., & Reeves, M. (2025, April 10). Understanding the global macroeconomic impacts of trump's tariffs. *Harvard Business Review*. https://hbr.org/2025/04/understanding-the-global-macroeconomic-impacts-of-trumps-tariffs

White House. (2025). *Fact sheet: President Donald J. Trump declares national emergency to increase our competitive edge, protect our sovereignty, and strengthen our national and economic security*. https://www.whitehouse.gov/fact-sheets/2025/04/fact-sheet-president-donald-j-trump-declares-national-emergency-to-increase-our-competitive-edge-protect-our-sovereignty-and-strengthen-our-national-and-economic-security/

World Meteorologic Organisation. (2025). *WMO confirms 2024 as warmest year on record at about 1.55°C above pre-industrial level*. World Meteorologic Organisation. https://wmo.int/news/media-centre/wmo-confirms-2024-warmest-year-record-about-155degc-above-pre-industrial-level

Chapter 2
The Economic Impact of US-China Tensions

Silvo Dajčman, Dejan Romih, and Valentin Toader

Abstract In this paper, we analyse the impact of a sudden and unexpected increase in US-China tensions on the US and Chinese economies using vector autoregression. We find that a positive shock in US-China tensions has a negative impact on both economies, although there are certain differences between them due to the characteristics of their economies and their ability to adapt to change. The findings indicate that a sudden and unforeseen rise in geopolitical frictions negatively influences industrial output and stock market performance in each country. However, the impact on exports presents a different picture—China appears to have greater flexibility in diversifying its global trade partnerships. The results of the study have policy implications as they emphasise the need to strengthen the resilience of both the US and Chinese economies to geopolitical shocks. This also applies to other economies.

Keywords China · Shock · Tension · United States

2.1 Introduction

Nowadays, the world economies are intensely interconnected, with countries trading goods, sharing technology, and relying on each other for resources. As a result, geopolitical tensions (conflicts, sanctions, political instability, etc.) can rapidly generate effects abroad; a disruption in one region can lead to global shortages, inflation, and market volatility in other countries. Thus, while being connected can bring benefits like growth and cooperation, it also means that problems spread more

S. Dajčman (✉) · D. Romih
Faculty of Economics and Business, University of Maribor, Maribor, Slovenia
e-mail: silvo.dajcman@um.si; dejan.romih@um.si

V. Toader
Faculty of Business, Babeş-Bolyai University, Cluj-Napoca, Romania
e-mail: valentin.toader@ubbcluj.ro

A. L. Negrusa, M. M. Coroş (eds.), *Innovation, Sustainability, and Growth in a VUCA Environment*, Springer Proceedings in Business and Economics,
https://doi.org/10.1007/978-3-032-11639-0_2

easily. As a result, countries are now trying to protect themselves better by spreading out their risks and being more careful about whom they depend on.

Geopolitical tensions can have a detrimental effect on economic, financial, and other parts of domestic and international society (Caldara & Iacoviello, 2022; Le et al., 2022; Jia et al., 2023; Jawadi et al., 2024). Geopolitical tensions between the United States and China, given the size, economic and political impact of the two economies, may not be detrimental only to their economies (Ding et al., 2022; Chor & Li, 2024; Rogers et al., 2024) but for the global economy and markets (Cheng et al., 2022; Cai, 2025).

Measuring bilateral tensions in real time is a challenging task that, for the US-China relation, has only recently been successfully resolved by Rogers et al. (2024), who utilise machine learning on the US newspaper text search related to tensions between the two countries.

This paper applies the US-China tensions index of Rogers et al. (2024) to study the effects of shocks in US-China tensions on the growth of industrial production, exports, stock prices, and inflation in the United States and China. We find that US-China tension shock negatively impacts the US and Chinese industrial production growth and stock market returns, while the overall exports are negatively impacted only in the United States.

2.2 Literature Review

In line with the previous studies (Caldara & Iacoviello, 2022; Le et al., 2022; Jawadi et al., 2024), for the purpose of the current study, we define geopolitical tensions as conflicts and risks (violent or non-violent), arising from political, military, economic, or social struggles between states or political actors. These tensions may involve war, terrorism, or other disruptions, including power rivalries and territorial disputes, that threaten or destabilise the normal, peaceful conduct of international relations.

The geopolitical tensions hinder economic growth by disrupting trade, deterring investment, and increasing uncertainty (Le et al., 2022; Jawadi et al., 2024). The effects are not only immediate but also long-lasting, continuing after the initial shock. Conflicts and tensions damage infrastructure, reduce private consumption and productivity, and limit governments' fiscal capacities. Looking at the Euro Area, Jawadi et al. (2024) point out the negative impact that a higher press coverage of tensions has on short-term GDP. Furthermore, they conclude that the impact is more sensitive to geopolitical threats than to acts. They highlight inflation and energy price volatility as key transmission channels, affecting global supply chains and trade flows, especially due to Europe's reliance on energy imports. Le, Bui, and Uddin (2022) analyse the conflicts' impact from the perspective of the income disparities between regions. Their findings show that low- and lower-middle-income countries experience more significant negative effects on economic growth and population well-being, especially in the long term. In contrast, high-income countries are more resilient.

Geopolitical tensions reduce investment, both in the short- and long-term, mainly because they create uncertainty and risk, which makes investors more cautious. Focusing on the United States, Caldara and Iacoviello (2022) find that increases in geopolitical risk (both threats and events) generate declines in investment, employment, and stock prices, which in the end negatively impact GDP growth. Conducting a firm-level analysis, they conclude that the industries which are more exposed to these risks suffer even larger investment drops, the companies deciding to cut investments in the long run. A similar impact can be observed in the case of FDIs. Aiyar, Malacrino, and Presbitero (2024) find that geopolitical distance (a measure of political misalignment) strongly reduces FDI flows: the more politically distant two countries are, the less likely they are to invest in each other, and this effect has become stronger over time (the change from a low to high level of geopolitical distance, from the 25th to the 75th percentile, is associated with a 15% drop in FDI). This impact is especially strong in low- and middle-income countries and in strategic sectors such as semiconductors. Since 2018, this pattern has grown stronger, especially due to the rising US-China tensions. Investments are increasingly concentrated between geopolitically aligned countries.

International trade is also affected by the rise of geopolitical tensions. Indirect effects also occur, as investors' sentiment is negatively affected by regional spillovers or changes in the global trade patterns. For example, the US-China trade war, one of the current sources of geopolitical tensions, impacts negatively, indirectly, the EU countries through supply chain disruptions, global uncertainty, and changes in trade patterns and prices. These tensions determine firms to diversify supply chains and influence GDP growth, investment, and stock markets. The media coverage of the trade conflict plays a strong role in shaping expectations, which in turn influences economic performance (Beckmann et al., 2024). Violent conflicts have an even more accentuated impact due to their spillover effects. Conflicts not only harm the country directly involved but can also spread instability to neighbouring regions, amplifying uncertainty, and disrupting regional and international trade. These disruptions are the result of increased risks, damaged infrastructure, and reduced trust in cross-border commerce (Le et al., 2022).

The literature confirms the negative impact geopolitical tensions have on financial markets; however, the magnitude and transmission channels vary by economic structure, market openness, and geographic context. Salachas et al. (2024) find that geopolitical shocks lead to significant decreases in stock returns in both developed and emerging markets, with the strongest effects in emerging economies due to lower financial resilience and higher exposure to foreign capital: a one-standard-deviation increase in the Geopolitical Risk index leads to an average monthly stock return decline of 0.77%, while in the case of emerging markets it leads to 0.83%. These shocks increase market volatility, drive investors toward safe-haven assets like gold, and are often worsened by other global crises, such as climate change and COVID-19. On the other hand, financial markets in the Euro Area appear to be less sensitive to geopolitical tensions due to economic diversification, global corporate reach, strong institutional responses, and a generally long-term investment outlook (Jawadi et al., 2024). Yilmazkuday (2024) supports the same idea, adding that the impact varies by the openness of the market and the proximity to the geopolitical event. According to his results, more open and globally integrated markets are more sensitive to geopolitical risks due to their dependence on global trade and investment flows. Also, he emphasises that the impact of the Russo-Ukrainian war had the biggest impact on neighbouring economies.

As mentioned above, precious metals, mainly gold and silver, are perceived as effective hedges against geopolitical risk, particularly in the short and medium term. Baur and Smales (2020) add that the price of precious metals reacts more strongly to geopolitical threats than to realised geopolitical acts. On the other hand, palladium and platinum serve only this role when geopolitical events directly disrupt their supply chains, affecting their industrial use (Cheng et al., 2022).

The US-China tensions are having a negative impact on the global economy. The general perspective is that China shows a higher economic vulnerability, the effects being asymmetrically localised within the country. Considering the size of both economies and their importance in international trade, the frictions between them have the capacity to adjust global supply chains and investment flows, most of the time to the detriment of global economic stability. Additionally, reciprocal actions increase uncertainty and increase economic loss. As a result, the economic impact of the US-China geopolitical tensions represents a subject intensely studied in the literature, the section below emphasises some of these results.

Regarding the strength of impact, Ding et al. (2022) concluded that while both countries experience welfare declines, China records higher welfare losses than the United States in the bilateral tariff war (−0.2391% vs. –0.0729%) taking place between 2018 and 2019. As a reaction to the US tariffs, China imposed retaliatory tariffs and strengthened its collaboration within the Asia-Pacific region by participating in the Regional Comprehensive Economic Partnership (RCEP). These actions allowed China to compensate for only a small part of the impact (11.92%), highlighting that even if they help, they cannot counterbalance the negative effects of bilateral frictions with a major economic partner. However, the negative effects are not equally distributed. The trade war reduces the overall real income in both countries, with some consumers being obliged to bear a higher part of the costs. For example, tariffs may have electoral motivations, targeting swing counties in the United States to hurt the regions that support the president imposing the tariffs more—in this case, retaliatory Chinese tariffs disproportionately impacted Republican-leaning agricultural areas, undercutting the political benefits of the trade war for the Republican Party (Fajgelbaum & Khandelwal, 2022). Similarly, the US tariffs targeted specific Chinese regions, 2.5% of China's population in the regions most exposed to tariffs saw income per capita fall by 2.52% and employment by 1.62% (Chor & Li, 2024).

Rogers et al. (2024) introduce the China-US tensions index to measure the extent of tension in the bilateral relations as perceived by the public through news in leading US newspapers. They show that the index closely matches the perceptions of the tensions by business and policy decision makers in the United States. They show that the higher the index, the lower the investments of US firms, especially for the more exposed US companies, and the greater the negative impact on US-China trade relations. Higher tensions also reflect in the stock market valuation of US companies. Fitting a vector autoregression model and impulse response function, they show that US-China tensions shocks depress the US and Chinese industrial production, exports of the United States to China and imports of China from the United States (more the latter than the former), appreciate USD against the yuan, and impact financing conditions of US firms as expressed by excess bond premium.

The economic impact of US-China tensions extends beyond these countries. Cai (2025) shows how trade restrictions and increased uncertainty generate worldwide costs, delay investments, and constrain global liquidity. Additionally, the disruption of global supply chains is encouraging companies to re-shore and diversify, especially in critical sectors like semiconductors, further strengthening industrial slowdowns and volatility in global trade. Similarly, Liu et al. (2025) conclude that US-China tensions determine both American and Chinese firms to reallocate FDIs to reduce geopolitical risk. US firms are diversifying their supply chains away from China, favouring politically stable emerging markets. Correspondingly, Chinese firms are shifting FDIs toward countries aligned with their strategic interests, such as those in the Belt and Road Initiative or markets offering indirect access to the US.

2.3 Research Methodology and Data

We analyse the impact of US-China tensions shocks by estimating impulse response functions based on the estimates of the vector autoregression (VAR) model. The VAR model consists of a constant vector term (c) and a vector of n endogenous variables (y_t), including the US-China tensions index ($TENSION_t$), share price growth ($SHAREgr_t$), industrial production growth ($IPgr_t$), export growth ($EXPgr_t$), and inflation (INF_t), or formally:

$$y_t = c + \sum_{l=1}^{q} N_l y_{t-l} + e_t \tag{2.1}$$

where q is determined by information criteria, N_l is a matrix of parameter estimates, and e_t is an error term vector.

Since the impact of the shock on China's and the US economy is expected to be different, we separately study each country and thus estimate country-specific VAR models and the resulting impulse responses. This differs from Rogers et al. (2024), who estimate one VAR model for the United States and China. Estimating separate country models also reduces the problem of shock identification. Shock identification in impulse response analysis is achieved by Cholesky decomposition, with the ordering of variables as given above, implying that the US-China tensions index does impact but is not impacted contemporaneously by share price growth and other variables in the model, share price growth is contemporaneously impacted by the US-China tensions index but not the other variables in the model, etc. This reasoning is identical to Baker et al. (2016) and differs from Rogers et al. (2024). On the ordering of other economic variables, see, e.g., Dajčman et al. (2023). The alternative ordering, similar to Rogers et al. (2024), is part of the robustness check. Our models differ from that of Rogers et al. (2024) also by a set of variables. We opt for a more parsimonious model and include only a comparable set of variables for both economies. Particularly, we include all exports of an investigated country, not only bilateral exports, since following the tensions shock, there might be an adjustment in the geographical structure of exports and thus the tension in bilateral relations may be detrimental only for bilateral exports but not the total exports of a country. All computations are performed with Stata.

The data sample consists of monthly observations. The period ranges 2010m1–2024m4, which is also different from Rogers et al. (2024), who focus on a 20-year period before December 2019. All data but the tensions index enter the model in growth rates, computed as: monthly observation$_t$/monthly observation$_{t-12}$ $* 100 - 100$, i.e., we observe the growth of a variable compared to the same month of the previous year. We opted for growth rates because the data on China's industrial production is available only in growth rates. Exports of the United States and China are to all foreign countries (because the data on Chinese exports to the United States is not available). Stationarity was not tested, as this reflects established practice in the field. Table 2.1 details the data series features and sources.

Figure 2.1 depicts the dynamics of all variables in the model.

2.4 Results and Discussions

The US-China tensions index reached the highest level in March 2020, coinciding with the outbreak of the COVID-19 pandemic and the first measures to contain the virus, and the United States blaming China for the outbreak (Rogers et al., 2024). The second-highest spike is observed in February 2022, which, among others, coincides with the US introduction of America COMPETES Act (China Briefing, 2025) and the Russia–Ukraine war (Rogers et al. 2024). Industrial production, exports, and share price growth visibly fell during the first period, but the association between the US-China tensions index and other variables is not clearly apparent during other periods.

Figure 2.2, which conveys the impulse responses of variables[1] to the US-China tensions index shock, shows that a positive shock to the US-China tensions index negatively affects economic outcomes in the observed economies.

[1] The VAR model (1) for the United States and China includes 1 lag ($q = 1$) as specified by SBIC information criteria. For the US model we also tested a model with first two lags ($q = 2$), not presented here, which does not qualitatively impact the results.

Table 2.1 Time series notation, description, and sources

Notation	Description and data source
$TENSION_t$	US-China tensions index, monthly frequency, untransformed series (Rogers et al., 2024). Data was retrieved from the Economic Policy Uncertainty data portal (https://www.policyuncertainty.com/US_China_Tension.html)
$IPgr\_US_t$	Industrial production growth rate, monthly frequency. Growth rate was computed as described above from the industrial production index (2017 = 100, seasonally adjusted) data from the Board of Governors of the Federal Reserve System (2025)
$IPgr\_C_t$	Industrial production growth rate, monthly frequency. Data was retrieved from the National Bureau of Statistics of China (2025). Note that for 2010 and 2011, there is no data for January, while for other years, there is no data for January and February. Gaps in sample data are treated by Stata's default setting, i.e., by omission of missing time entries
$SHAREgr\_US_t$	Share price growth rate for the US, monthly frequency. Growth was computed as described above from index series (Index 2015 = 100, Monthly, Not Seasonally Adjusted) data obtained from Organization for Economic Cooperation and Development (2025a)
$SHAREgr\_C_t$	Share price growth rate for the US, monthly frequency. Growth was computed as described above from index series (Index 2015 = 100, Monthly, Not Seasonally Adjusted) data obtained from Organization for Economic Cooperation and Development (2025b)
$EXPgr\_US_t$	Exports of goods and services for the US, seasonally adjusted. The growth rate was computed as described above, based on data obtained from the US Census Bureau and the US Bureau of Economic Analysis (2025)
$EXPgr\_C_t$	Export of commodities for China, converted to US Dollar, seasonally adjusted. The growth rate was computed as described above, based on data obtained from the Organization for Economic Cooperation and Development (2025c)
INF_US_t	Inflation for the United States, monthly growth rate, computed from the consumer price index (1982–1984 = 100, seasonally adjusted); data obtained from the US Bureau of Labor Statistics (2025)
INF_C_t	Inflation for China, growth rate, computed as described above from consumer price index (2015 = 100, not seasonally adjusted); data obtained from the Organization for Economic Cooperation and Development (2025d)

Source: Own

Export growth in the United States experiences a statistically significant decline following the tensions index shock. The strongest response (a decrease in the growth rate by 1.6 percentage points) is observed 3 months after the shock. The Chinese export growth is not significantly impacted by the shock. Industrial production growth reduces significantly in both countries, more in the United States (the largest response is 3 months after the shock by approximately 0.9 percentage points reduction) than in China. The stock market also significantly responds to tensions between the United States and China: the share price growth (returns) responds negatively on impact in the United States and China, more in the latter than in the former. Clearly, the stock prices respond to the US-China tensions shock more quickly than exports and industrial production and, at least in the United States, also recover more quickly than the real economic variables. Inflation dynamics are not significantly impacted by the tension shock.

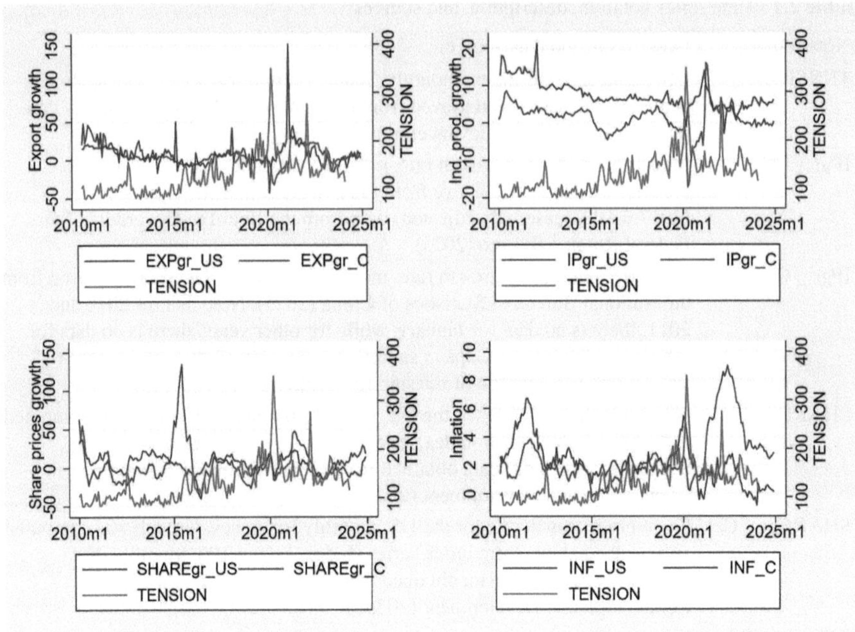

Fig. 2.1 The US-China tensions index and the growth of industrial production, export, share prices, and inflation for the United States and China. (Notes: EXPgr_US, IPgr_US, SHAREgr_US, and INF_US are time series for the US, while EXPgr_C, IPgr_C, SHAREgr_C, and INF_C are time series for China. Source: Rogers et al. (2024), Board of Governors of the Federal Reserve System (2025), National Bureau of Statistics of China (2025), Organization for Economic Cooperation and Development (2025a, b, c, d), US Census Bureau and US Bureau of Economic Analysis (2025), and US Bureau of Labor Statistics (2025))

In addition to applying a different lag of the baseline model for the United States (see footnote 1), the robustness of the results is also checked by applying a different ordering of variables (the lag order of VARs remains 1 as in the baseline specification). Following Rogers et al. (2024), IPgr_t was placed first, followed by INF_t, EXPgr_t, TENSION_t, and SHAREgr_t, but the baseline results are robust to this specification (the impulse responses to a one standard deviation shock to TENSION_t are presented in Fig. 2.3).

The presented results show that increased tensions in the US-China relations undermine the economic performance of the countries involved, which corroborates the findings from the literature. The dampening effect on economic activity and trade is larger for the United States than for China, which seems surprising and may show that China can economically adjust to shocks in its relations with the United States more easily than the United States can, which is in line with findings from Rogers et al. (2024). While these authors show that bilateral exports to the United States drop upon the US-China tensions index, our results show that the impact on overall exports cannot be confirmed, which further indicates the adaptability of the Chinese economy to the shock.

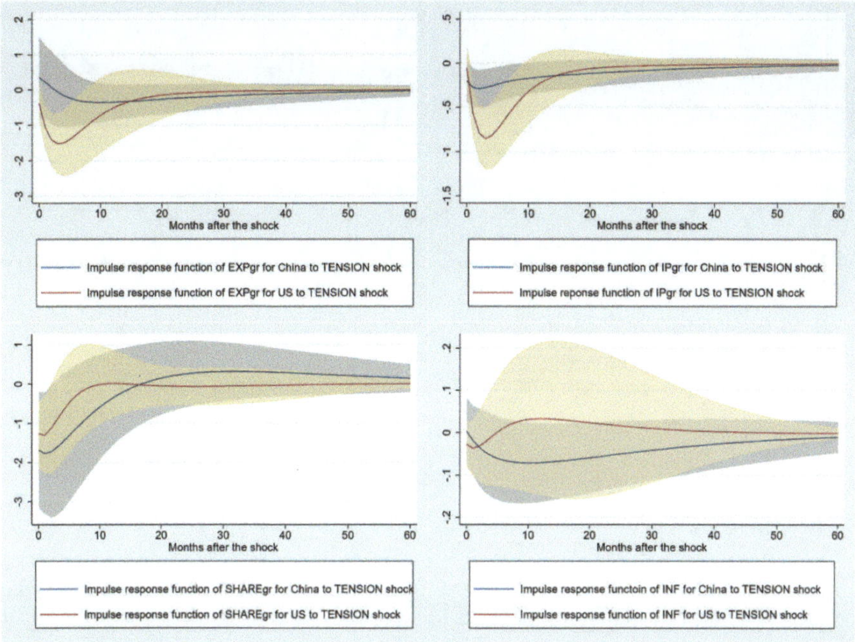

Fig. 2.2 Impulse response functions to a shock in the US-China tensions index. (Note: The figure presents orthogonal impulse responses to a one standard deviation shock in the US-China tensions index. The responses (Y axis) are in percentage points. The shaded area shows the 95% confidence intervals. Source: Own calculations)

2.5 Conclusions

The aim of this paper was to analyse the impact of US-China tensions on the United States and Chinese economies. The results show that a sudden and unexpected increase in US-China tensions has a negative impact on the growth of industrial production and stock prices in both countries. The situation is different for exports, as China has more room to diversify its international trade. This means that geopolitics is not insignificant, but its economic consequences depend on various factors, in particular the ability of countries to compensate for a fall in foreign demand.

The results of the study are valuable for researchers and policymakers as they emphasise the importance of managing geopolitical risks and reducing tensions between countries. The analysis focused on two large economies; however, future research could usefully extend this work to smaller economies.

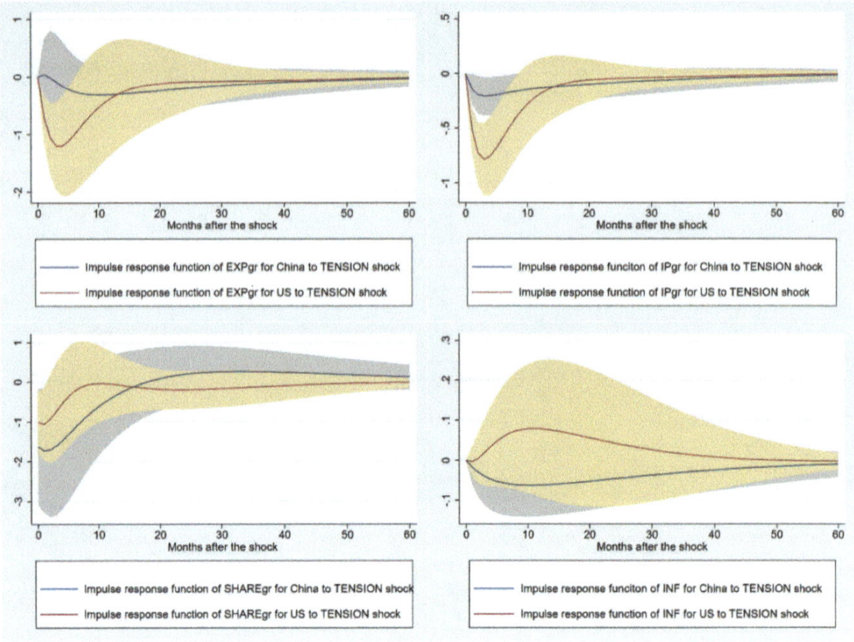

Fig. 2.3 Impulse response functions to a shock in the US-China tensions index under alternative ordering of variables. (Notes: The notes from Fig. 2.2 apply. Source: Own calculations)

Appendix

Information criteria results for the optimal lag selection—VAR model for the US

Lag-order selection criteria

Sample: 2010m5 thru 2024m2 Number of obs = 166

Lag	LL	LR	df	p	FPE	AIC	HQIC	SBIC
0	-2684.34				8.1e+07	32.4017	32.4398	32.4955
1	-1969.59	1429.5	25	0.000	19975.3	24.0915	24.3198	24.6539*
2	-1914.35	110.5	25	0.000	13886*	23.7271*	24.1456*	24.7581
3	-1893.28	42.138*	25	0.017	14589.1	23.7744	24.3832	25.2742
4	-1877.25	32.053	25	0.156	16316.8	23.8825	24.6815	25.851

* optimal lag
Endogenous: TENSION SHAREgr_US IPgr_US EXPgr_US INF_US
Exogenous: _cons

Information criteria results for the optimal lag selection—VAR model for China

Lag-order selection criteria

Sample: 2010m6 thru 2023m12, but with gaps Number of obs = 87

Lag	LL	LR	df	p	FPE	AIC	HQIC	SBIC
0	-1439.63				1.8e+08	33.2098	33.2669	33.3516
1	-1082.42	714.41	25	0.000	87957.6	25.5729	25.9153*	26.4232*
2	-1054.51	55.812*	25	0.000	82744.1*	25.5061*	26.1338	27.065
3	-1039.01	31.004	25	0.189	104424	25.7244	26.6375	27.9919
4	-1021.79	34.45	25	0.099	128332	25.9032	27.1016	28.8793

* optimal lag
Endogenous: TENSION SHAREgr_C IPgr_C EXPgr_C INF_C
Exogenous: _cons

References

Aiyar, S., Malacrino, D., & Presbitero, A. (2024). Investing in friends: The role of geopolitical alignment in FDI flows. *European Journal of Political Economy, 83*, 102508. https://doi.org/10.1016/j.ejpoleco.2024.102508

Baker, S. R., Bloom, N., & Davis, S. J. (2016). Measuring economic policy uncertainty. *The Quarterly Journal of Economics, 131*(4), 1593–1636. https://doi.org/10.1093/qje/qjw024

Baur, D., & Smales, L. (2020). Hedging geopolitical risk with precious metals. *Journal of Banking & Finance, 117*, 105823. https://doi.org/10.1016/j.jbankfin.2020.105823

Beckmann, J., Czudaj, R., & Murach, M. (2024). Macroeconomic effects from media coverage of the China–US trade war on selected EU countries. *European Journal of Political Economy, 85*, 102611. https://doi.org/10.1016/j.ejpoleco.2024.102611

Board of Governors of the Federal Reserve System (US). (2025). *Industrial production: Total Index [INDPRO]*. FRED, Federal Reserve Bank of St. Louis. Retrieved February 21, 2025, from https://fred.stlouisfed.org/series/INDPRO

Cai, Y. (2025). US-China tensions, global supply chains pressure, and global economy. *Economics Letters, 250*, 112283. https://doi.org/10.1016/j.econlet.2025.112283

Caldara, D., & Iacoviello, M. (2022). Measuring geopolitical risk. *American Economic Review, 112*(4), 1194–1225. https://doi.org/10.1257/aer.20191823

Cheng, S., Zhang, Z., & Cao, Y. (2022). Can precious metals hedge geopolitical risk? Fresh sight using wavelet coherence analysis. *Resources Policy, 79*, 102972. https://doi.org/10.1016/j.resourpol.2022.102972

China Briefing. (2025). *US-China relations in the Biden era: A timeline*. https://www.china-briefing.com/news/US-China-relations-in-the-biden-era-a-timeline/

Chor, D., & Li, B. (2024). Illuminating the effects of the US-China tariff war on China's economy. *Journal of International Economics, 150*, 103926. https://doi.org/10.1016/j.jinteco.2024.103926

Dajčman, S., Kavkler, A., Levenko, N., & Romih, D. (2023). Spillover effects of economic policy uncertainty on adult and youth unemployment. *Review of Economic Perspectives, 23*(1), 47–70. https://doi.org/10.2478/revecp-2023-0001

Ding, G., Guoa, G., Wua, C., & Yu, J. (2022). China-US trade friction and welfare: The role of regional trade agreements. *Economic Modelling, 113*, 105877. https://doi.org/10.1016/j.econmod.2022.105877

Fajgelbaum, P., & Khandelwal, A. (2022). The economic impacts of the US-China trade war. *Annual Review of Economics, 14*, 205–228. https://doi.org/10.1146/annurev-economics-051420-110410

Jawadi, F., Rozin, P., Gnegne, Y., & Cheffou, A. (2024). Geopolitical risks and business fluctuations in Europe: A sectorial analysis. *European Journal of Political Economy, 85*, 102585.

Jia, R., Roberts, M. E., Yang, Y., & Yang, E. (2023). The imapct of US-China tensions on US science: Evidence from the NIH investigations. *Proceedings of the National Academy of Sciences of the United States of America, 121*(19), e2301436121. https://doi.org/10.1073/pnas.2301436121

Le, T.-H., Bui, M.-T., & Uddin, G. (2022). Economic and social impacts of conflict: A cross-country analysis. *Economic Modelling, 115*, 105980.

Liu, L., Gozgor, G., Mahalik, M., & Pal, S. (2025). The impact of the US-China tensions on FDI dynamics in emerging economies. *Finance Research Letters, 78*, 107255. https://doi.org/10.1016/j.frl.2025.107255

National Bureau of Statistics of China. (2025). *Value-added of industry, growth rate (The same period last year=100)(%)*. Retrieved from https://data.stats.gov.cn/english/easyquery.htm?cn=A01

Organization for Economic Cooperation and Development. (2025a). *Financial market: Share prices for United States* (SPASTT01USM661N). FRED, Federal Reserve Bank of St. Louis. Retrieved February 21, 2025, from https://fred.stlouisfed.org/series/SPASTT01USM661N

Organization for Economic Cooperation and Development. (2025b). *Financial market: Share prices for China* (SPASTT01CNM661N). FRED, Federal Reserve Bank of St. Louis. Retrieved February 21, 2025, from https://fred.stlouisfed.org/series/SPASTT01CNM661N

Organization for Economic Cooperation and Development. (2025c). *International merchandise trade statistics: Exports: Commodities for China* (XTEXVA01CNM667S). FRED, Federal Reserve Bank of St. Louis. Retrieved February 21, 2025, from https://fred.stlouisfed.org/series/XTEXVA01CNM667S

Organization for Economic Cooperation and Development. (2025d). *Consumer price indices (CPIs, HICPs), COICOP 1999: Consumer Price Index: Total for China* (CHNCPIALLMINMEI). FRED, Federal Reserve Bank of St. Louis. Retrieved February 21, 2025, from https://fred.stlouisfed.org/series/CHNCPIALLMINMEI

Rogers, J. H., Sun, B., & Sun T. (2024, May 3). US-China tension. https://doi.org/10.2139/ssrn.4815838

Salachas, E., Kouretas, G., Laopodis, N., & Vlamis, P. (2024). Stock market spillovers of global risks and hedging opportunities. *European Journal of Political Economy, 83*, 102533. https://doi.org/10.1016/j.ejpoleco.2024.102533

US Bureau of Labor Statistics. (2025). *Consumer Price Index for all urban consumers: All items in US city average* (CPIAUCSL). FRED, Federal Reserve Bank of St. Louis. Retrieved February 21, 2025, from https://fred.stlouisfed.org/series/CPIAUCSL

US Census Bureau and US Bureau of Economic Analysis. (2025). *Exports of goods and services, balance of payments basis* (BOPTEXP). FRED, Federal Reserve Bank of St. Louis. Retrieved February 21, from https://fred.stlouisfed.org/series/BOPTEXP

Yilmazkuday, H. (2024). Geopolitical risk and stock prices. *European Journal of Political Economy, 83*, 102553. https://doi.org/10.1016/j.ejpoleco.2024.102553

[illegible]

Chapter 3
Transforming Workplace Well-Being Through Innovation and Embodiment

Sébastien Point and Elisabeta Butoi

Abstract In the context of escalating workplace stress, digital hyperconnectivity, and emerging technological disruptions, this chapter reimagines the concept of workplace well-being by moving beyond traditional cognitive and managerialist framings. Drawing from both classical models—subjective, eudaimonic, and affective well-being—and emerging perspectives, the chapter emphasizes the embodied, relational, and technological dimensions of well-being. It explores three critical frontiers: the tension between quantified and quantum paradigms of well-being; digital disconnection as a restorative practice amid hyperconnectivity; and the neurosomatic imperative, which re-centers the body as a site of intelligence, healing, and relational atonement. Through this multidimensional lens, the chapter critiques instrumental approaches to well-being and proposes an integrated framework that includes artificial and neurosomatic intelligences. It concludes with a research agenda for the hospitality industry, urging scholars and practitioners to explore how these new trajectories can reshape guest experiences and employee training.

Keywords Workplace well-being · Eudaimonic well-being · Neurosomatic intelligence · Quantum well-being · Digital disconnection · Artificial intelligence · Hospitality industry · Embodiment

> The body never lies. (Martha Graham, American modern dancer, teacher, and choreographer)

S. Point (✉)
EM Strasbourg Business School, Université de Strasbourg, Strasbourg, France
e-mail: point@unistra.fr

E. Butoi
Faculty of Business, Babeş-Bolyai University, Cluj-Napoca, Romania
e-mail: elisabeta.butoi@ubbcluj.ro

A. L. Negrusa, M. M. Coroş (eds.), *Innovation, Sustainability, and Growth in a VUCA Environment*, Springer Proceedings in Business and Economics,
https://doi.org/10.1007/978-3-032-11639-0_3

3.1 Introduction: The Urgency of Well-Being in the Modern Workplace

As organizations navigate a volatile, uncertain, complex, and ambiguous (VUCA) environment, attention to employee well-being has become not only a moral imperative but a strategic one (Kowalski & Loretto, 2017). This shift gained urgency during the COVID-19 pandemic, which acted as a global stress test for organizational resilience and individual mental health (Kim et al., 2024). For instance, Kim, Han, and Ariza-Montes (2021) show that guests place greater importance on well-being-related hotel attributes when forming brand loyalty. As far as employees are concerned, Mihalache and Mihalache (2022) demonstrate that organizational and supervisory support significantly buffers stressors such as overload and uncertainty to improve job-related well-being and affective commitment. Wellness in hospitality is shaped by high job demands, emotional labor, irregular schedules, and organizational support systems (Alcorn & Koo, 2025). As an illustration, companies like Accor have recently launched creative initiatives to enhance well-being.[1]

Well-being at work is increasingly recognized as a multidimensional construct that encompasses physical, psychological, emotional, social, and spiritual facets of an employee's experience. Rather than being confined to the absence of illness or stress, this holistic approach frames well-being as a dynamic interplay of individual, relational, and organizational factors. Physical well-being pertains to safety, ergonomics, and physical health management, while psychological and emotional well-being relate to stress regulation, resilience, and positive affect—often shaped by leadership style, job design, and perceived autonomy (Skakon et al., 2010; Nielsen & Daniels, 2012). Social well-being emphasizes the role of collegial support, inclusion, and recognition in enhancing trust and reducing isolation, with evidence showing that positive work climates and value alignment promote satisfaction and engagement (Kristof-Brown et al., 2005). Spiritual well-being, though less frequently studied, has emerged as a vital dimension linked to purpose, meaning, and ethical alignment at work—especially in contexts where employees seek coherence between personal values and organizational missions (Dale & Burrell, 2014; Eva et al., 2019). Together, these dimensions call for an integrated model of well-being that transcends narrow performance metrics and recognizes employees as whole, embodied individuals.

Therefore, workplace well-being is not merely an outcome to be managed or optimized—it is an experience to be lived, questioned, and redefined. This chapter investigates the evolving nature of workplace well-being, exploring both its measurement and embodied experience. To return to the opening quote, we draw from Martha Graham's radical insight to foreground an underexplored but powerful thesis in organizational behavior: well-being is inherently embodied. This lens moves us beyond cognitive models and metrics, prompting us to consider the

[1]Accor, The Future of Work: How Accor Prioritizes Talent Well-Being, published on-line 20/01/2025, https://group.accor.com/en/Actualites/2025/01/future-of-work-talent-well-being

neurophysiological, emotional, and existential dimensions of work. This approach echoes the concept of "organized embodiment," which highlights how organizational practices impose specific physical and emotional demands on employees and the need to address the embodied realities of employees within organizational contexts (Dale & Burrell, 2014).

Recently, Saito, Brozović, and Baum (2025) present a systematic review of 175 peer-reviewed articles examining the multidimensional construct of well-being among hospitality employees, categorizing it across physical, psychological, and social dimensions. Their analysis reveals substantial gaps in the existing literature—particularly a lack of attention to structural inequalities, marginalized worker experiences, and the impact of emerging technologies. Their work implicitly opens a critical avenue for linking employee well-being with developments in Artificial Intelligence (AI). In the same vein, Wong et al. (2024) highlight the need for future research to address emerging technological influences—including AI—on employee mental health and psychological well-being in hospitality and tourism. Consequently, this chapter aims to explore more contemporary trends and innovations in the conceptualization of well-being, with a particular emphasis on the intersection of artificial and neurosomatic intelligence. By doing so, we seek to re-center the body in well-being discourses, considering both its quantification and its potential transformation through quantum-informed approaches.

This chapter adopts a conceptual and exploratory approach. Rather than presenting empirical data or testing hypotheses, it synthesizes existing literature across disciplines—including organizational psychology, hospitality studies, neuroscience, and technology studies—to propose a multidimensional framework for workplace well-being. The research investigates the balance between regulation and innovation, questioning the effects of AI on service delivery and employee engagement. It also explores how self-quantification impacts well-being, focusing on behavioral change and awareness, further examining the role of wearable technologies in enhancing employee well-being.

The structure of this chapter unfolds in two main parts, each addressing distinct yet interconnected dimensions of workplace well-being. The first part critically revisits the classical domains of workplace well-being, encompassing psychological, physical, and social aspects, while simultaneously engaging with the darker undercurrents that permeate organizational life—namely, stress, burnout, emotional exhaustion, and toxic workplace cultures. The second part turns to three emergent and increasingly salient trajectories that challenge and expand conventional understandings of well-being. These include the tension between quantified and quantum approaches to well-being; this tension highlights the shift from data-driven self-optimization to more holistic, entangled understandings of health and consciousness. In conclusion, this chapter invites readers to move beyond narrow, instrumentalized conceptions of well-being. Instead, it advocates for a rethinking of workplace well-being as a lived, relational, and embodied phenomenon.

3.2 Mapping the Landscape of Workplace Well-Being

Initially focused on health and safety, the early discourse on the so-called "well-being" prioritized protection from physical risks in industrial settings. During the 1970s–1980s, attention shifted to psychological distress, introducing terms like stress, fatigue, and burnout to describe the mental toll of modern work. By the 1990s, the focus broadened to include employee engagement and happiness, emphasizing motivation, job satisfaction, and affective responses to work. It is only in the 2010s that the term "well-being" gained formal and strategic prominence in organizational contexts, coinciding with a broader multidimensional approach that integrates mental health, emotional resilience, and systemic factors such as leadership and organizational culture.

This historical arc demonstrates that well-being has often been framed in contrast to its absences—be it physical safety, emotional strain, or disengagement—thus affirming that the study of well-being is frequently intertwined with its opposites.

Classical Dimensions of Well-Being

The literature traditionally distinguishes between subjective, eudaimonic, and affective well-being. Subjective well-being refers to individuals' assessments of their lives, integrating both cognitive evaluations, such as life satisfaction and emotional experiences, including the frequency of positive and negative feelings. The theory posits that individuals are more likely to report higher well-being when they possess resources that align with their personal values and life goals (Diener et al., 1999). Rooted in the field of happiness economics, the concept underlines the potential for public policy to improve societal well-being by considering individuals' self-assessed happiness and life satisfaction. In organizations, higher subjective well-being is linked to improved productivity, better job performance, and lower absenteeism (Bryson et al., 2017). Supervisor and coworker support enhances well-being, while stressors from work demands diminish it (Monnot & Beehr, 2014). Despite extensive research, a universally accepted definition of subjective well-being remains elusive, suggesting that assessments of life satisfaction and happiness can vary significantly and encompass both cognitive evaluations and emotional experiences (Zheng, 2024).

Eudaimonic well-being centers on the realization of personal potential, meaningful engagement, and living in accordance with one's true self. Unlike hedonic well-being, which emphasizes pleasure and the avoidance of discomfort, eudaimonic well-being reflects a deeper sense of purpose, self-actualization, and psychological flourishing. Increasingly acknowledged as a critical dimension of overall well-being, it offers unique insights into optimal psychosocial functioning and the quality of life. Research by Joshanloo (2018) underscores the empirical distinctiveness of eudaimonic well-being from subjective well-being, revealing its moderate positive association with national indicators such as life satisfaction and gross domestic product. Moreover, individual-level factors—particularly positive affect and educational attainment—emerge as significant predictors of eudaimonic well-being.

The construct is shaped by both individual attributes and sociocultural conditions, affirming its multifaceted nature. Complementing this, Jacob and Reddy (2024) demonstrate that social cognitive skills training can enhance adolescents' capacity to discover purpose and meaning in life. Their findings highlight the potential of targeted interventions to foster mental health and social development through the cultivation of eudaimonic capacities.

Mindfulness plays a pivotal role in fostering eudaimonic well-being by cultivating enhanced attention, present-moment awareness, and non-judgmental acceptance (Kay et al., 2017). These core elements of mindfulness promote positive psychological traits—such as self-acceptance and emotional resilience—while attenuating negative affective responses to stress and adversity. According to He and Liu (2023), mindfulness-based interventions have demonstrated efficacy in enhancing eudaimonic well-being across both clinical and non-clinical populations, underscoring their value in supporting personal growth, meaning-making, and sustained psychological health. In the Hospitality industry, Zhang, Cheng, and Ma (2024) show that tourism experiences can elicit eudaimonic well-being by fostering meaning, autonomy, and personal growth, as revealed through online travel reviews.

An illustration of the distinction between hedonic and eudaimonic well-being is offered by Rahmani, Gnoth, and Mather (2018), who conducted a psycholinguistic analysis of a large corpus of travel blogs using text topic modeling techniques. Their findings revealed that hedonic well-being was frequently articulated through language rich in positive affect—such as terms like "amazing," "beautiful," and "fun"—which signify ephemeral pleasure and sensory enjoyment. In contrast, expressions of eudaimonic well-being were more prevalent in narratives reflecting transformative and introspective experiences, including accounts of cultural immersion, personal challenge, and self-reflection. Such narratives were characterized by purpose-driven and meaning-oriented language, underscoring the role of travel in facilitating personal growth and existential insight. In the same vein, Huang, Wang, and Wu (2024) propose an integrative framework linking transformative tourism experiences to both hedonic and eudaimonic well-being. This study underscores the potential of transformative tourism to foster both immediate pleasure and long-term personal development.

Affective well-being refers to the balance of positive and negative emotions that individuals encounter in response to daily life events. As a central component of subjective well-being, it plays a significant role in shaping both mental and physical health outcomes (Ishio & Abe, 2017). Various studies have explored different aspects of affective well-being, including its measurement, its relationship with life satisfaction (Berlin & Fors Connolly, 2019), and its implications across diverse life domains such as workplaces, religious involvement, and physical activity engagement. Within organizational settings, affective well-being has been identified as a key determinant of overall employee happiness and job satisfaction, with clear implications for performance and retention. For instance, leaders' stress levels, emotional states, and supportive behaviors are significantly associated with employee affective outcomes—both positive and negative. In particular, transformational and supportive leadership styles tend to enhance employees' emotional well-being (Skakon et al., 2010). Conceptually, affective well-being is informed by a tripartite framework comprising dispositional, activity-based, and telic (goal-oriented) sources of affective states. This framework helps elucidate the interplay of personality traits, job characteristics, and emotional intelligence in shaping emotional responses at work. Notably, effective coping strategies and the capacity for recovery from occupational stress are strongly correlated with heightened levels of positive affect and reduced negative emotionality (Hosie & Sevastos, 2010). Methodologically, affective well-being is often assessed using tools like the Day Reconstruction Method (DRM), which captures momentary affect through participants' retrospective accounts of daily episodes. Recent innovations have integrated DRM with physiological stress indicators, such as heart rate variability measured via wearable devices, suggesting promising avenues for objective proxies of affective states (Ishio & Abe, 2017). Daniels (2000) also identifies five key bipolar dimensions of affective well-being at work, such as anxiety–comfort and tiredness–vigor. His framework highlights the multifaceted nature of workplace emotions and provides tools for assessing employee affect.

Workplace well-being is often most vividly understood through the lens of its absence. In organizational research, the study of well-being is frequently anchored in its opposites—stress, burnout, emotional exhaustion, anxiety, and other manifestations of impaired mental health. For instance, emotional labor and role overload significantly reduce job satisfaction and health, underscoring the need for proactive stress management in hotel HRM practices (O'Neill & Davis, 2011). Figure 3.1

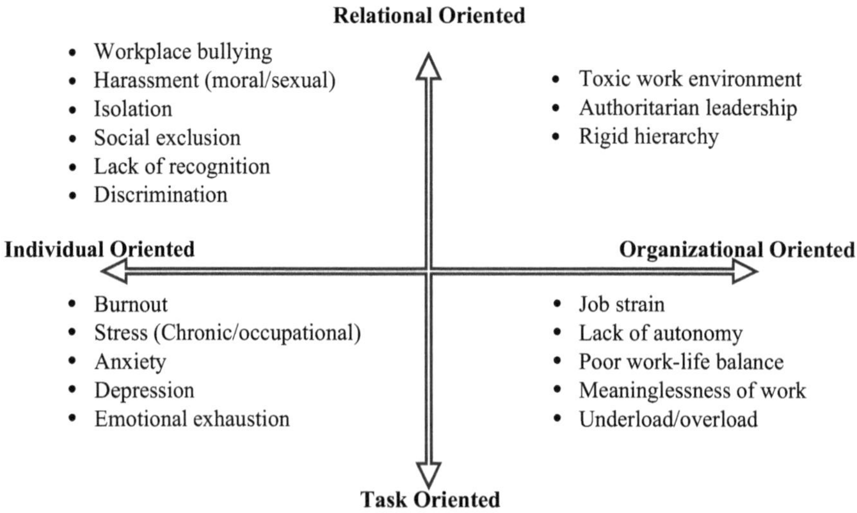

Fig. 3.1 Antecedents and outcomes of impaired well-being. (Source: Authors)

underscores this perspective by mapping the multi-dimensional drivers of non-well-being across individual, task, organizational, and relational domains. It illustrates that diminished well-being is not merely the result of personal vulnerability but is shaped by systemic dynamics such as poor job design, authoritarian leadership, lack of autonomy, workplace bullying, and social exclusion. These factors are not isolated; they interact and compound, revealing that well-being must be seen as a relational, contextual, and embodied phenomenon.

Cultivating Well-Being at Work: The Roles of Leadership, Job Characteristics, Creativity, and Organizational Culture

Empirical research has consistently demonstrated that well-designed jobs contribute to higher levels of psychological engagement, job satisfaction, and overall well-being (Humphrey et al., 2007). For instance, enriched job characteristics are well-established predictors of psychological well-being (Nielsen & Daniels, 2012). Equally, the behavior and style of leadership significantly affect employees' affective states as leadership has a measurable impact on employee stress and well-being (Skakon et al., 2010).

Transformational leadership, characterized by individualized consideration and inspirational motivation, has been found to enhance employee affective well-being and reduce emotional exhaustion by fostering trust (Kelloway et al., 2012). Transformational leadership positively influences employees' perceptions of work characteristics—such as autonomy, clarity, and support—which in turn enhance their psychological well-being (Nielsen et al., 2008). Furthermore, Liu, Siu, and Shi (2010) show that transformational leaders enhance well-being by building interpersonal trust and fostering employees' belief in their own capabilities. This suggests that affective well-being at work is shaped not only by external leadership behaviors but also by the internal psychological resources that such leadership cultivates in followers (Nielsen & Munir, 2009). In the hospitality industry, transformational leadership significantly enhances employee well-being by improving the quality of work life, which in turn reduces burnout, boosts organizational commitment, and increases overall life satisfaction (Kara et al., 2013).

The Servant Leadership style positively influences workplace well-being (Eva et al., 2019), which in turn enhances employee creativity. The presence of psychological availability further strengthens this relationship, suggesting that supportive leadership can foster a conducive environment for creative engagement (Song et al., 2024). Creative efforts aimed at making work more enjoyable are associated with higher life satisfaction and positive emotions. This suggests that creativity can be a direct source of personal happiness and meaning, beyond material benefits (Sherman & Shavit, 2017). Helzer and Kim (2019) argue that creativity is a key driver of workplace well-being, fostering meaning, engagement, and positive emotions. They highlight how creative expression can buffer stress and enhance psychological resilience in organizational settings.

Finally, organizational culture plays a crucial role in shaping employee well-being and the overall workplace environment. Research has shown that supportive and inclusive cultures foster psychological safety, trust, and positive affective states, thereby enhancing both subjective and eudaimonic well-being (Kuenzi & Schminke, 2009). Moreover, the alignment between organizational culture and employees' personal values can serve as a key predictor of job satisfaction and emotional well-being (Kristof-Brown et al., 2005). Conversely, toxic or misaligned cultures—marked by rigidity, hierarchy, or hostility—can undermine well-being by increasing role stress, burnout, and disengagement (Schneider et al., 2013). Therefore, implementing a culture of support is crucial for ensuring employee well-being. As such, cultivating a well-being-oriented culture is not merely a matter of ethics or morale but a strategic imperative for sustainable organizational performance.

3.3 Innovation and Integration: New Frontiers in Well-Being

As the landscape of work continues to evolve under the influence of technological disruption, shifting social expectations, and a renewed emphasis on human-centered values, new paradigms are emerging that challenge traditional models of workplace well-being.

Quantified Versus Quantum Well-Being

Recent advances introduce novel paradigms such as quantified self-practices and quantum well-being. Quantified well-being focuses on the use of self-tracking technologies to monitor and improve personal health metrics, while quantum well-being explores the potential of quantum phenomena in nutrition and health. Both approaches offer unique insights and applications, yet they differ fundamentally in their methodologies and implications for well-being.

Self-quantification refers to the practice of systematically tracking personal health data—such as physical activity, emotional states, and social interactions—through digital tools like wearables and mobile applications. This practice is central to the Quantified Self (QS) movement, which aspires to promote self-knowledge and personal optimization by leveraging digital data (Jain et al., 2024). Empirical studies suggest that self-quantification can contribute positively to individual well-being by fostering behavioral change and enhancing awareness of one's physiological and psychological states (Feng et al., 2021). However, scholars also caution that the pervasive monitoring encouraged by QS technologies may foster self-objectification, body image anxiety, and diminished self-esteem, particularly in individualistic and performance-driven cultural contexts (Jain et al., 2024; De Moya & Pallud, 2020). More precisely, self-quantification can lead to self-objectification, potentially harming well-being by shifting individuals' focus from intrinsic experiences to external metrics and appearance-based evaluations (Peterson Fronczek et al. 2022).

While direct studies on self-quantification within the hospitality industry are limited, several relevant works explore the integration of wearable technologies and data-driven personalization to enhance employee well-being and guest experiences. For instance, Troisi, Visvizi, and Grimaldi (2023) highlight how digitalization and data-driven business models in hospitality ecosystems can enable self-quantification practices that personalize guest experiences and optimize service delivery through real-time data analytics. Else, Mejia, Ciarlante, and Chheda (2021) propose a wearable technology solution to improve housekeeper safety and health, illustrating how self-quantification through real-time tracking of physical activity and environmental conditions can enhance occupational well-being in the hospitality industry.

Quantum well-being is an emerging interdisciplinary concept that integrates principles from quantum physics with aspects of health, nutrition, and consciousness to enhance overall well-being. Quantum technology, including quantum sensing, computing, and communication, is poised to enhance community services such as healthcare, education, and environmental conservation. These technologies promise improved precision in diagnostics and resource allocation, contributing to community well-being (Phillipson, 2024). The relationship between quantum physics and consciousness suggests that quantum principles can influence higher brain functions and emotional regulation. In organizations, quantum computing has the potential to solve certain types of problems—especially in optimization, simulation—and developing a quantum computing strategy today is not about immediate ROI, but about positioning a company to benefit from tomorrow's technological breakthroughs (Bova et al., 2023).

While the term "quantum well-being" is not explicitly used in hospitality literature, several studies and industry insights explore the integration of quantum principles and advanced technologies to enhance guest experiences and well-being. For instance, Adhikari and Chang (2021) investigate the impact of quantum computing on supply chain management (SCM) and hotel performance, emphasizing its potential to transform operational processes in the hospitality sector. The study highlights how quantum computing enables real-time data processing, complex optimization, and advanced predictive analytics, leading to more accurate demand forecasting, inventory control, and dynamic pricing strategies.

The contrast between "doing" and "being" approaches to well-being parallels the emerging distinction between quantified and quantum paradigms of well-being—where the former emphasizes measurable performance and self-optimization, while the latter invokes a more holistic, interconnected, and experiential understanding of health and consciousness. This distinction between "doing" and "being" approaches to well-being captures a fundamental tension in contemporary organizational life: whether well-being is treated as a set of externally driven behaviors and performance outcomes, or as an intrinsically valued, lived experience (see Table 3.1). The "doing" approach tends to align with instrumental, productivity-focused models of well-being, where interventions such as resilience training, mindfulness apps, or employee wellness programs are evaluated based on their return on investment or

Table 3.1 "Doing" vs. "Being" approaches to well-being

	"Doing" mode (instrumental, data-driven)	"Being" mode (experiential, spiritual)
Ontology	Human as a measurable, optimizable resource	Human as a conscious, evolving, meaning-making being
Practices	Bio-tracking (wearables), dashboards, KPI's	Mindfulness, introspection
Metrics	Engagement scores, HR analytics, sleep data	Subjective meaning, spiritual fulfillment, sense of purpose
Time orientation	Future-oriented: optimize for results	Present-centered: awareness and presence
Leadership model	Transformational leadership	Servant authentic or spiritual leadership
Culture	Lean, data-informed, metrics-based culture	Psychologically safe, human-centric culture
Risks/ Limitations	Surveillance, reductionism, loss of human complexity	Vagueness, lack of measurability, resistance to institutional integration
Keywords	Efficiency, productivity, biohacking, neuroscience, AI, quantified self	Presence, self-transcendence, inner life, values, consciousness, meaning

Source: Authors

performance enhancement. In contrast, the "being" approach emphasizes well-being as an embodied, relational, and existential state, grounded in authenticity, self-awareness, and value congruence. This resonates with emerging critiques in organizational theory that argue for a shift from managerialist metrics to human-centered and ethically grounded paradigms (Dale & Burrell, 2014).

The convergence of quantified-self technologies and quantum computing with artificial intelligence represents a paradigm shift in digital innovation, enabling hyper-personalized insights and exponentially more powerful data processing that can redefine health, decision-making, and human–machine interaction. AI and algorithms are transforming workplaces with both potential benefits and risks. The main challenge is that current labor laws weren't designed for AI systems and may be inadequate for regulating algorithmic management in workplaces.

This creates tension between two approaches: strict regulation to protect workers versus minimal rules to encourage innovation. IA has been intensely investigated in the hospitality industry. For instance, Ruel and Njoku (2021) explore how AI is reshaping the hospitality industry by transforming service delivery, operational efficiency, and workforce dynamics. They develop a theoretical framework that highlights AI's impact on employee engagement, customer experience, and strategic decision-making. Law et al. (2024) have recently conducted a systematic review of 100 studies on artificial intelligence in hospitality, highlighting a strong focus on service robots and customer-facing applications, while identifying gaps in mid-service operations and employee perspectives. Furthermore, Dwivedi et al. (2024) examine the emerging role of generative AI tools like ChatGPT in the hospitality and tourism industry, outlining their applications in customer service (e.g., automated guest interactions and virtual concierges), marketing (e.g., personalized promotional content and chatbot-driven engagement), and content creation (e.g., real-time itinerary generation and review summarization). They emphasize how these technologies can enhance operational efficiency, personalization, and customer satisfaction, while also transforming traditional service delivery models.

Digital Disconnection and the Neurosomatic Intelligence

Hyper-connectivity through Internet of Things (IoT) technologies enhances hospitality service delivery by enabling real-time personalization, operational efficiency, and guest satisfaction, while also addressing challenges related to data security, infrastructure demands, and guest privacy in increasingly connected environments (Mercan et al., 2021). Digital hyperconnectivity has become a silent epidemic, associated with cognitive fatigue, sleep disturbances, and emotional erosion. Policy responses, such as France's 2017 "droit à la déconnexion," institutionalize the right to disconnect outside of work hours (Mazmanian et al., 2013). Digital disconnection, the practice of intentionally reducing or limiting digital media use, is increasingly recognized as a strategy to enhance well-being. The effectiveness of digital disconnection varies depending on individual motivations, strategies employed, and contextual factors. Many individuals feel that their digital media use is out of balance, prompting them to disconnect to recalibrate their usage.

The effectiveness of disconnection strategies can vary based on individual motivations, industry-specific factors, and the presence of supportive policies and practices. Hassan and Saleh (2024) systematically review the emerging field of digital detox and digital-free tourism, identifying key motivations such as stress reduction, mindfulness, and reconnection with nature. Recognizing the harm of "always-on" cultures, some hotel brands enforce right-to-disconnect policies. For example, management avoids sending non-urgent messages after hours, and employees are discouraged from checking email during leave.

Recent papers discuss the balance between technological advancement and the need for human-centric experiences in hospitality (Ruel & Njoku, 2021). Given the implications of constant connectivity, the authors suggest that deliberate digital disconnection can enhance employee engagement and customer satisfaction. Consequently, digital disconnection is not anti-technology—it is pro-human ecology. It refers to the concept of the "neurosomatic intelligence," which refers to the essential role that the integration of neurological and somatic (bodily) health plays in achieving overall well-being and productivity in the workplace. Implementing neurosomatic principles can lead to improved emotional regulation, reduced anxiety, and enhanced physical health.

By focusing on the body's sensory experiences and integrating them with cognitive processes, individuals can achieve a more balanced and responsive nervous system. While direct references to neurosomatic intelligence in the hospitality sector are limited, the industry's emphasis on guest well-being and personalized experiences presents opportunities for application. For instance, the hospitality industry's emphasis on personalized guest experiences aligns with neurosomatic principles by addressing individual needs and promoting relaxation. Incorporating neurosomatic practices can enhance employee training programs, improve customer service interactions, and foster environments that promote relaxation and stress reduction for guests.

3.4 Conclusion: Humanizing the Future of Work

Given the conceptual nature of this chapter, its propositions are not intended as empirically validated conclusions but as theoretically grounded pathways for future exploration. The arguments are derived from interdisciplinary literature and are offered as a stimulus for further empirical validation in hospitality contexts.

This chapter has argued for a paradigmatic shift in how we understand and enact well-being within contemporary organizational life. Moving beyond metrics and managerialist paradigms, we have re-situated well-being as a lived, relational, and somatically grounded phenomenon. Drawing from both classical and emerging frameworks, we have demonstrated how well-being encompasses far more than affective or cognitive evaluations—it is fundamentally embodied, shaped by leadership dynamics, organizational cultures, job design, and increasingly, by the interplay of technology and corporeality.

In this light, we proposed three critical frontiers that redefine the boundaries of workplace well-being. First, we explored the continuum from quantified to quantum approaches, revealing a tension between data-centric optimization and a more holistic, entangled understanding of human flourishing. Second, we emphasized digital disconnection as a necessary countermeasure to hyperconnectivity, pointing to its potential for restoring attention, regulating emotions, and sustaining human agency. Third, we introduced the neurosomatic intelligence, which centers the body not as an object of productivity, but as a site of wisdom, healing, and relational intelligence—crucial to navigating the psychosocial demands of contemporary work.

As challenges, current labor laws are inadequate for regulating AI systems, creating tension between worker protection and innovation encouragement, as well as pervasive monitoring from self-quantification technologies may lead to self-objectification and diminished self-esteem. Overall, these limited studies on self-quantification in the hospitality industry highlight gaps in understanding employee perspectives.

Looking ahead, future research in the hospitality industry should engage more critically and empirically with these emerging paradigms. This includes examining how AI-driven personalization and quantum-informed service design affect both guest experience and employee well-being; exploring the implementation and outcomes of digital disconnection policies across diverse hospitality contexts; and investigating how neurosomatic practices might be operationalized in training, leadership, and wellness programs.

References

Adhikari, B., & Chang, B. Y. (2021). Quantum computing impact on SCM and hotel performance. *International Journal of Internet, Broadcasting and Communication, 13*(2), 1–6. https://doi.org/10.7236/IJIBC.2021.13.2.1

Alcorn, M., & Koo, B. (2025). Wellness dimensions of hospitality employees: A systematic review. *International Journal of Hospitality Management, 124*, 103958. https://doi.org/10.1016/j.ijhm.2024.103958

Berlin, M., & Fors Connolly, F. (2019). The association between life satisfaction and affective well-being. *Journal of Economic Psychology, 73*, 34–51. https://doi.org/10.1016/j.joep.2019.04.010

Bova, F., Goldfarb, A., & Melko, R. (2023). The business case for quantum computing. *MIT Sloan Management Review, 64*(3), 31–37.

Bryson, A., Forth, J., & Stokes, L. (2017). Does employees' subjective well-being affect workplace performance? *Human Relations, 70*(8), 1017–1037. https://doi.org/10.1177/0018726717693073

Dale, K., & Burrell, G. (2014). Being occupied: An embodied re-reading of organizational 'wellness'. *Organization, 21*(2), 159–177. https://doi.org/10.1177/1350508412473865

Daniels, K. (2000). Measures of five aspects of affective well-being at work. *Human Relations, 53*(2), 275–294. https://doi.org/10.1177/a010564

De Moya, J. F., & Pallud, J. (2020). From panopticon to heautopticon: A new form of surveillance introduced by quantified-self practices. *Information Systems Journal, 30*(6), 940–976. https://doi.org/10.1111/isj.12284

Diener, E., Suh, E. M., Lucas, R. E., & Smith, H. L. (1999). Subjective well-being: Three decades of progress. *Psychological Bulletin, 125*(2), 276. https://doi.org/10.1037/0033-2909.125.2.276

Dwivedi, Y. K., Pandey, N., Currie, W., & Micu, A. (2024). Leveraging ChatGPT and other generative artificial intelligence (AI)-based applications in the hospitality and tourism industry: Practices, challenges and research agenda. *International Journal of Contemporary Hospitality Management, 36*(1), 1–12. https://doi.org/10.1108/IJCHM-05-2023-0686

Eva, N., Robin, M., Sendjaya, S., Van Dierendonck, D., & Liden, R. C. (2019). Servant leadership: A systematic review and call for future research. *The Leadership Quarterly, 30*(1), 111–132. https://doi.org/10.1016/j.leaqua.2018.07.004

Feng, S., Mäntymäki, M., Dhir, A., & Salmela, H. (2021). How self-tracking and the quantified self promote health and well-being: Systematic review. *Journal of Medical Internet Research, 23*(9), e25171. https://doi.org/10.2196/25171

Hassan, T., & Saleh, M. I. (2024). Tourism digital detox and digital-free tourism: What do we know? What do we not know? Where should we be heading? *Journal of Tourism Futures*. https://doi.org/10.1108/JTF-12-2023-0274

He, Q., & Liu, X. (2023). The effects of mindfulness on eudaimonic well-being and its theoretical explanation. *Advances in Psychological Science, 31*(10), 1937. https://doi.org/10.3724/SP.J.1042.2023.01937

Helzer, E. G., & Kim, S. H. (2019). Creativity for workplace well-being. *Academy of Management Perspectives, 33*(2), 134–147. https://doi.org/10.5465/amp.2016.0141

Hosie, P. J., & Sevastos, P. P. (2010). A framework for conceiving of job-related affective wellbeing. *Management Revue, 21*(4), 406–436. https://doi.org/10.5771/0935-9915-2010-4-406

Huang, X., Wang, P., & Wu, L. (2024). Well-being through transformation: An integrative framework of transformative tourism experiences and hedonic versus eudaimonic well-being. *Journal of Travel Research, 63*(4), 974–994. https://doi.org/10.1177/00472875231171670

Humphrey, S. E., Nahrgang, J. D., & Morgeson, F. P. (2007). Integrating motivational, social, and contextual work design features: A meta-analytic summary and theoretical extension of the work design literature. *Journal of Applied Psychology, 92*(5), 1332. https://doi.org/10.1037/0021-9010.92.5.1332

Ishio, J., & Abe, N. (2017). Measuring affective well-being by the combination of the day reconstruction method and a wearable device: Case study of an aging and depopulating community in Japan. *Augmented Human Research, 2*(1), 2. https://doi.org/10.1007/s41133-017-0006-2

Jacob, L., & Reddy, K. J. (2024). Unlocking eudaimonic well-being: Assessing the impact of adolescent social cognitive skills training. In *Principles and clinical interventions in social cognition* (pp. 368–375). IGI Global.

Jain, A., Pradhan, D., Kuanr, A., & Singh, S. (2024). Self-quantification and consumer Well-being: A meta-analytic review. *Psychology & Marketing, 42*(2), 514–538. https://doi.org/10.1002/mar.22141

Joshanloo, M. (2018). Optimal human functioning around the world: A new index of eudaimonic well-being in 166 nations. *British Journal of Psychology, 109*(4), 637–655. https://doi.org/10.1111/bjop.12316

Kara, D., Uysal, M., Sirgy, M. J., & Lee, G. (2013). The effects of leadership style on employee well-being in hospitality. *International Journal of Hospitality Management, 34*, 9–18. https://doi.org/10.1016/j.ijhm.2013.02.001

Kay, A. A., Hafenbrack, A., & Skarlicki, D. (2017). Enhancing eudaimonic well-being with mindfulness: The moderating effect of authenticity. *Academy of Management Proceedings, 2017*, 14366. https://doi.org/10.5465/ambpp.2017.14366

Kelloway, E. K., Turner, N., Barling, J., & Loughlin, C. (2012). Transformational leadership and employee psychological well-being: The mediating role of employee trust in leadership. *Work & Stress, 26*(1), 39–55. https://doi.org/10.1080/02678373.2012.660774

Kim, J. J., Han, H., & Ariza-Montes, A. (2021). The impact of hotel attributes, well-being perception, and attitudes on brand loyalty: Examining the moderating role of COVID-19 pandemic. *Journal of Retailing and Consumer Services, 62*, 102634. https://doi.org/10.1016/j.jretconser.2021.102634

Kim, J., Lee, H. W., & Chung, G. H. (2024). Organizational resilience: Leadership, operational and individual responses to the COVID-19 pandemic. *Journal of Organizational Change Management, 37*(1), 92–115. https://doi.org/10.1108/JOCM-05-2023-0160

Kowalski, T. H., & Loretto, W. (2017). Well-being and HRM in the changing workplace. *The International Journal of Human Resource Management, 28*(16), 2229–2255. https://doi.org/10.1080/09585192.2017.1345205

Kristof-Brown, A. L., Zimmerman, R. D., & Johnson, E. C. (2005). Consequences of individuals' fit at work: A meta-analysis of person–job, person–organization, person–group, and person–supervisor fit. *Personnel Psychology, 58*(2), 281–342. https://doi.org/10.1111/j.1744-6570.2005.00672.x

Kuenzi, M., & Schminke, M. (2009). Assembling fragments into a lens: A review, critique, and proposed research agenda for the organizational work climate literature. *Journal of Management, 35*(3), 634–717. https://doi.org/10.1177/0149206308330559

Law, R., Lin, K. J., Ye, H., & Fong, D. K. C. (2024). Artificial intelligence research in hospitality: A state-of-the-art review and future directions. *International Journal of Contemporary Hospitality Management, 36*(6), 2049–2068. https://doi.org/10.1108/IJCHM-02-2023-0189

Liu, J., Siu, O. L., & Shi, K. (2010). Transformational leadership and employee well-being: The mediating role of trust in the leader and self-efficacy. *Applied Psychology, 59*(3), 454–479. https://doi.org/10.1111/j.1464-0597.2009.00407.x

Mazmanian, M., Orlikowski, W. J., & Yates, J. (2013). The autonomy paradox: The implications of mobile email devices for knowledge professionals. *Organization Science, 24*(5), 1337–1357. https://doi.org/10.1287/orsc.1120.0806

Mejia, C., Ciarlante, K., & Chheda, K. (2021). A wearable technology solution and research agenda for housekeeper safety and health. *International Journal of Contemporary Hospitality Management, 33*(10), 3223–3255. https://doi.org/10.1108/IJCHM-01-2021-0102

Mercan, S., Cain, L., Akkaya, K., Cebe, M., Uluagac, S., Alonso, M., & Cobanoglu, C. (2021). Improving the service industry with hyper-connectivity: IoT in hospitality. *International*

Journal of Contemporary Hospitality Management, 33(1), 243–262. https://doi.org/10.1108/IJCHM-06-2020-0621

Mihalache, M., & Mihalache, O. R. (2022). How workplace support for the COVID-19 pandemic and personality traits affect changes in employees' affective commitment to the organization and job-related well-being. *Human Resource Management, 61*(3), 295–314. https://doi.org/10.1002/hrm.22082

Monnot, M. J., & Beehr, T. A. (2014). Subjective well-being at work: Disentangling source effects of stress and support on enthusiasm, contentment, and meaningfulness. *Journal of Vocational Behavior, 85*(2), 204–218. https://doi.org/10.1016/j.jvb.2014.07.005

Nielsen, K., & Daniels, K. (2012). Does shared and differentiated transformational leadership predict followers' working conditions and well-being? *The Leadership Quarterly, 23*(3), 383–397. https://doi.org/10.1016/j.leaqua.2011.09.001

Nielsen, K., & Munir, F. (2009). How do transformational leaders influence followers' affective well-being? Exploring the mediating role of self-efficacy. *Work & Stress, 23*(4), 313–329. https://doi.org/10.1080/02678370903385106

Nielsen, K., Randall, R., Yarker, J., & Brenner, S. O. (2008). The effects of transformational leadership on followers' perceived work characteristics and psychological well-being: A longitudinal study. *Work & Stress, 22*(1), 16–32. https://doi.org/10.1080/02678370801979430

O'Neill, J. W., & Davis, K. (2011). Work stress and Well-being in the hotel industry. *International Journal of Hospitality Management, 30*(2), 385–390. https://doi.org/10.1016/j.ijhm.2010.07.007

Peterson Fronczek, L., Mende, M., & Scott, M. L. (2022). From self-quantification to self-objectification? Framework and research agenda on consequences for well-being. *Journal of Consumer Affairs, 56*(3), 1356–1374. https://doi.org/10.1111/joca.12478

Phillipson, F. (2024). Leveraging quantum technology to enhance community services and supportive ICT infrastructure. In *Innovations for community services* (pp. 36–47). Springer. https://doi.org/10.1007/978-3-031-60433-1_3

Rahmani, K., Gnoth, J., & Mather, D. (2018). Hedonic and eudaimonic well-being: A psycholinguistic view. *Tourism Management, 69*, 155–166. https://doi.org/10.1016/j.tourman.2018.06.008

Ruel, H., & Njoku, E. (2021). AI redefining the hospitality industry. *Journal of Tourism Futures, 7*(1), 53–66. https://doi.org/10.1108/JTF-03-2020-0032

Saito, H., Brozović, D., & Baum, T. (2025). Well-being of hospitality employees: A systematic literature review. *International Journal of Hospitality Management, 124*, 103955. https://doi.org/10.1016/j.ijhm.2024.103955

Schneider, B., Ehrhart, M. G., & Macey, W. H. (2013). Organizational climate and culture. *Annual Review of Psychology, 64*(1), 361–388. https://doi.org/10.1146/annurev-psych-113011-143809

Sherman, A., & Shavit, T. (2017). The thrill of creative effort at work: An empirical study on work, creative effort and well-being [dataset]. In *PsycEXTRA dataset*. American Psychological Association (APA). https://doi.org/10.1037/e506002017-001

Skakon, J., Nielsen, K., Borg, V., & Guzman, J. (2010). Are leaders' well-being, behaviours and style associated with the affective well-being of their employees? A systematic review of three decades of research. *Work & Stress, 24*(2), 107–139. https://doi.org/10.1080/02678373.2010.495262

Song, W., Ma, Y., Fan, X., Jin, X., & Peng, X. (2024). Servant leadership, workplace well-being and employee creativity: The roles of psychological availability and experienced creative time pressure. *Creativity and Innovation Management, 33*(3), 399–413. https://doi.org/10.1111/caim.12595

Troisi, O., Visvizi, A., & Grimaldi, M. (2023). Digitalizing business models in hospitality ecosystems: Toward data-driven innovation. *European Journal of Innovation Management, 26*(7), 242–277. https://doi.org/10.1108/EJIM-09-2022-0540

Wong, A. K. F., Kim, S., Gamor, E., Koseoglu, M. A., & Liu, Y. (2024). Advancing employees' mental health and psychological well-being research in hospitality and tourism: Systematic review, critical reflections, and future prospects. *Journal of Hospitality & Tourism Research, 49*, 5. https://doi.org/10.1177/10963480241271084

Zhang, X., Cheng, L., & Ma, G. (2024). Eliciting eudaimonic well-being in the tourism experiential space: Evidence from online reviews. *Tourism Management, 105*, 104955. https://doi.org/10.1016/j.tourman.2024.104955

Zheng, W. (2024). Analyse subjective wellbeing- based on definition, measurement, and limitations. *Advances in Economics, Management and Political Sciences, 75*(1), 205–209. https://doi.org/10.54254/2754-1169/75/20241668

Chapter 4
The Role of Higher Education in Facilitating Labor Market Integration

Teodora Cajvan and Daniela-Mihaela Neamțu

Abstract This study investigates the role of higher education in improving the employability of young graduates in the North-East region of Romania, highlighting its contribution to bridging the gap between academic training and labor market demands. The research aims to identify the determinants of graduates' success in securing and maintaining employment relevant to their fields of study, while analyzing discrepancies between academic curricula and employer expectations.

Using a mixed-methods approach, the study combines quantitative and qualitative methodologies to provide comprehensive insights. A structured survey was conducted with 200 graduates across disciplines and 11 employers in the region. Survey questions explored educational experiences, job search strategies, and the applicability of skills in professional settings. The data were analyzed using the Statistical Package for the Social Sciences (SPSS), employing Principal Component Analysis (PCA) to identify key factors influencing employability. These determinants include career advancement opportunities, salary competitiveness, and practical application of academic training. Additionally, semi-structured interviews with employers offered nuanced perspectives on essential competencies such as communication skills, technical expertise, digital literacy, and adaptability.

Results indicate that 82.5% of graduates secured employment post-graduation, but 60% were employed outside their academic specialization, underscoring a mismatch between training and labor market needs. Employers emphasized the importance of internships, university-business collaborations, and experiential learning to better prepare graduates for professional challenges.

This study underscores the need for higher education institutions to align curricula with labor market requirements by fostering university-employer partnerships, incorporating practical learning experiences, and developing transversal skills.

T. Cajvan (✉) · D.-M. Neamțu
Universitatea "Ştefan cel Mare" din Suceava, Suceava, Romania
e-mail: teodora.cajvan@student.usv.ro; dana.neamtu@usm.ro

A. L. Negrusa, M. M. Coroş (eds.), *Innovation, Sustainability, and Growth in a VUCA Environment*, Springer Proceedings in Business and Economics,
https://doi.org/10.1007/978-3-032-11639-0_4

Enhancing career counseling services and promoting entrepreneurship and lifelong learning are essential for equipping graduates with the competencies needed for sustained career success in a competitive labor market.

Keywords Higher education · Labor force · Human capital · Insertion

4.1 Introduction

In a world marked by rapid change driven by technological advancement, globalization, and structural economic transformations, access to and the quality of higher education represent fundamental pillars in the process of individual and societal development. In this regard, higher education is no longer perceived merely as the final stage of formal learning, but rather as a complex process of building human capital—a strategic resource for sustainable economic growth and for adapting to the increasingly dynamic demands of the labor market.

Public and academic discourse increasingly highlights a disconnection between academic training and the practical realities of employment. This gap is reflected in the challenges faced by young graduates in integrating into the labor market, in the high incidence of employment outside their field of specialization, and in employers' often critical perception of the new generation's preparedness for the workforce. At the same time, universities—constrained by accreditation standards and under pressure to deliver measurable outcomes—tend to prioritize theoretical knowledge at the expense of practical application, generating structural imbalances between educational supply and labor market demand.

In this context, a fundamental rethinking of the university's role is necessary—beyond its traditional mission of intellectual formation. Universities must become strategic partners in regional and national economic development by designing flexible educational programs aligned with private sector needs, supporting innovation and entrepreneurship, and cultivating transversal skills that transcend a single field of study. Critical thinking, adaptability, digital competencies, effective communication, and teamwork are becoming sine qua non conditions for successful professional integration.

This study addresses these issues through an applied investigation focused on the specific context of the North-East region of Romania—an area marked by economic vulnerabilities but also by significant potential in sectors such as IT, tourism, education, and services. The research aims to assess the extent to which university training influences graduate employability, how partnerships between academia and the business sector facilitate the transition to employment, and which competencies are most valued by employers during recruitment.

Using a mixed-methods approach—combining quantitative analysis (via a questionnaire applied to a sample of 200 graduates) and qualitative research (semi-structured interviews with 11 regional employers)—this study provides a comprehensive, multidimensional perspective on the relationship between education and professional integration. Based on statistical analyses (including correlations and principal component analysis—PCA), key factors contributing to professional success are identified: practical experience (through internships or part-time jobs), digital and social skills, field of specialization, and the prestige of the academic institution.

Thus, the study contributes to the academic literature by offering robust empirical evidence regarding the effectiveness of university training in relation to actual labor market requirements. Furthermore, the results may serve as a foundation for better-targeted educational policies, curricular adaptation, and strengthened cooperation mechanisms between universities and employers.

This paper ultimately reflects the urgent need to align higher education with contemporary economic demands in an integrated effort to transform academic institutions into active catalysts for the personal and professional development of young people. Only through a collaborative approach—supported by rigorous research and continuous dialogue with the socio-economic environment—can the education system fulfill its fundamental role in building an equitable, innovative, and sustainable society.

4.2 Literature Review

Higher education, defined as an advanced level of instruction, plays a critical role in preparing young individuals for the labor market (Brennan, 2010) and in fostering global collaboration for societal advancement (Cortese, 2003). Universities significantly contribute to the development of human capital, which encompasses the skills and knowledge essential for socio-economic success (Janice et al., 2016). Human capital, a foundational element of economic development, is driven by innovation and education and has a substantial impact on organizational performance (Mascaradi et al., 2022; Neamtu & Burac, 2023).

Although critics of human capital theory highlight its limitations in fully explaining the linkage between education and skill development (Jabłoński, 2021), the theory remains pivotal in exploring the relationship between education, labor market outcomes, and income levels (Buta et al., 2010). Schultz (1960) introduced the concept of "human wealth," illustrating the connection between education and productivity growth. Reder and Becker (1967) distinguished between general and specific human capital, expanding the application of the theory into labor economics and health (Teiceira, 2014). Mincer (1958) investigated the relationship between educational attainment and income distribution, establishing a theoretical framework that remains influential.

To foster sustainable economic growth, a balanced investment in both the demand for education and the supply of skilled labor is essential, supported by a synergy between market dynamics and social policy interventions (Schultz, 1968). The analysis of graduates' integration into the labor market provides valuable insights into aligning education with national economic needs.

The labor market—central to economic discourse—describes the interaction between labor supply and demand, influencing wages and employment conditions

(Eldridge, 2011). Its structure is shaped by worker preferences and macroeconomic forces, affecting job allocation and compensation patterns (DiPrete & Nonnemaker, 1997). In the context of digitalization, highly skilled workers with expertise in emerging technologies drive workplace transformation and operational efficiency (de Assis Dornelles et al., 2022).

Wage disparities across regions and industries reflect the value and demand for various skill sets. In Romania, sectors such as IT and research offer high remuneration due to their strategic importance and labor market demand, whereas low-value-added fields experience lower wage levels. Therefore, aligning educational outcomes with employer expectations is critical to ensuring effective labor market integration (Dimian, 2014; Patache, 2011).

Investments in education and the development of human capital remain essential for achieving sustainable economic growth and promoting equity in labor market participation (Institutul Naţional de Statistică, 2023). Development strategies must prioritize high-potential industries and facilitate access to education to strengthen the link between economic demands and professional training.

Econometric models play a central role in labor market research, providing robust tools to understand the interplay between economic and social variables. Mincer's model (1974) remains a seminal reference for analyzing the returns to education, examining the association between years of schooling and wage outcomes. Heckman (1979) introduced a model that corrects selection bias and which enhances the accuracy of employment and unemployment studies.

The North-East region of Romania is characterized by pronounced economic and social disparities, being among the least developed in the country. Studies on tourism in this area reveal significant cultural and natural potential, yet also underline challenges related to infrastructure and promotion. Arionesei et al. (2016) conducted a comparative analysis of tourism indicators using statistical techniques such as ANOVA to evaluate regional differences.

Tourism's impact on local economies and societies is another area of interest in the literature. Gaube (2015) emphasized the importance of sustainable tourism development, outlining both positive and negative effects on local economies and the environment. Other studies, such as that of Şuşu (2007), explore the tourism potential of counties like Bacău, Neamţ, and Suceava, focusing on cultural and religious heritage as well as regional development initiatives.

4.3 Research Methodology

Research Aim, Objectives, and Hypotheses

The primary aim of this research is to analyze how higher education contributes to the labor market integration of graduates in the North-East region of Romania. The study seeks to identify correlations and key determinants of professional success using data processed through SPSS.

Research Objectives:

- **O1.** To assess the relationship between the graduates' field of study and their employment outcomes.
- **O2.** To identify the influence of educational level and acquired competencies on labor market integration.
- **O3.** To analyze the impact of university-employer partnerships on employment opportunities.
- **O4.** To investigate the time required to secure employment based on individual strategies.
- **O5.** To explore employer perceptions regarding graduate preparedness and alignment with labor market demands.

Research Hypotheses:

- **H1.** Fields of study aligned with labor market demands significantly increase the probability of rapid employment.
- **H2.** Competencies acquired during academic studies positively influence employment success.
- **H3.** Partnerships between higher education institutions and employers have a significant effect on graduate employment rates.
- **H4.** The duration of the job search varies depending on individual strategies and professional domains.
- **H5.** Significant discrepancies exist between employer perceptions and competencies acquired through university programs.

The study will investigate these hypotheses through descriptive statistics, association tests, and regression analyses to gain a comprehensive understanding of graduate employment outcomes.

Research Instrument

The quantitative component of the study examines how higher education institutions in Romania's North-East region influence graduate students' employability. Drawing on responses from 200 graduates, the research identifies key factors that shape labor market adaptation. The study employs a mixed-methods approach, combining quantitative data from a structured questionnaire with qualitative insights from interviews.

The target population consists of 200 graduates from the region. The main objective is to assess the influence of higher education on successful labor market integration, identifying essential factors contributing to graduates' professional success.

The questionnaire was pre-tested on a subgroup of 30 respondents to ensure clarity and validity. The final version includes 25 questions, grouped into six distinct sections:

1. **Socio-demographic profile** (4 items): age, gender, university attended, and education level.
2. **Academic-professional alignment** (3 items): specialization versus current occupation, and graduate employment status.

3. **Job-seeking strategies** (3 items): approaches used to find employment, types of contracts, and job search duration.
4. **Labor market perspectives** (6 items): recruitment process, market evolution, university-employer collaborations, and skills developed during studies.
5. **Skills and employability** (4 items): skills gained, identified gaps, complexity of labor market integration, and guidance received.
6. **Educational expectations and future prospects** (5 items): academic program evaluation, labor market opportunities, and emerging professions.

In the qualitative component, a semi-structured interview guide was designed to explore employers' perspectives on graduate employability. This format facilitates the collection of detailed, context-sensitive information, focusing on respondents' perceptions and professional experiences rather than quantitative data (Ugwu & Eze, 2023).

The interviews were conducted online via a dedicated link and distributed through email. Data collection took place between February and March. The sample consisted of 11 employers from the North-East region of Romania, operating in diverse sectors including real estate, financial services, IT, and other industries.

The interview guide included 17 questions, structured into five thematic sections:

1. **Organizational profile (5 items):** company name, location, field of activity, year of establishment, and number of employees.
2. **Skills and professional preparation (3 items):** key competencies sought in graduates and expectations regarding academic training.
3. **Career orientation and work experience (4 items):** preferred education level, pathways for gaining professional experience, and advice for young graduates.
4. **Labor market requirements (3 items):** workforce retention strategies, projected changes in employment demand, and attractive future professions.
5. **Career transitions (2 items):** reasons for job changes among young employees and employer attitudes toward internships and student engagement.

The analysis aimed to compare employers' insights with graduates' experiences, offering a comprehensive understanding of the transition from higher education to the labor market.

4.4 Processing and Interpretation of Results

The Importance of Correlation Analysis in Studying the Determinants of Graduate Employability

Statistical analysis plays a fundamental role in educational and socio-economic research, providing in-depth insights into the relationships between key variables that influence graduates' transition from academia to the labor market. Correlation analysis, in particular, serves as a powerful tool for identifying interdependencies among factors such as practical experience, acquired competencies, university reputation, and the employability level of graduates. The inclusion of a correlation matrix in such analyses enhances the interpretability of statistical relationships, supporting the formulation of rigorous conclusions and guiding further research directions (Field, 2018).

Pearson's correlation coefficient is one of the most widely used measures to assess the strength and direction of linear associations between variables, offering reliable indicators of significance typically at the 0.01 (2-tailed) level (Hair et al., 2019). For instance, a strong correlation between internship experience and success in recruitment processes indicates that employers place high value on practical experience during candidate evaluation (Cohen et al., 2013).

This method of statistical analysis is crucial for identifying variables with a major influence on professional insertion, providing a clearer understanding of labor market requirements. The identification of strong correlations between digital and technical skills and early career success suggests that contemporary industries prioritize candidates with advanced technological competencies. This trend underscores the need for curriculum reform to align educational outcomes with current labor market demands (Tabachnick & Fidell, 2019).

Furthermore, correlation analysis contributes to the optimization of recruitment strategies by highlighting the criteria that employers deem essential. A significant correlation between social skills and employment success illustrates the growing recognition of soft skills across diverse professional sectors. This finding emphasizes the importance of preparing students for effective interpersonal interaction and collaboration in workplace settings (Bryman & Cramer, 2020).

Thus, the inclusion of a correlation table is not merely a methodological necessity but also a critical tool for validating conclusions and identifying the key variables that shape the professional trajectories of graduates. Statistical analysis, therefore, becomes a cornerstone of employability research, offering a structured framework for evaluating the educational impact on labor market integration.

Significant Correlations Between Educational Variables and Graduate Labor Market Integration

In the context of a knowledge-based economy, the effective integration of graduates into the labor market has become a strategic priority for both higher education institutions and policymakers (Tomlinson, 2017). The relevance of academic training, the development of transversal competencies, and practical experiences during university studies are crucial elements that directly influence graduate employability (Harvey, 2003; Yorke, 2006). Furthermore, the formation of human capital through higher education is associated with more efficient labor market outcomes and higher economic returns for both individuals and society (Becker, 1993; Mincer, 1974).

This analysis focuses on the most statistically significant correlations ($p \leq 0.001$) identified through a quantitative study conducted in the North-East region of Romania. The results highlight the relationships between educational background, acquired skills, and graduate integration into the labor market.

Key Findings

- Marketing specialization and career flexibility ($r = 0.815$, $p \leq 0.001$). Graduates in the field of Marketing benefit from high adaptability in the labor market, with access to a wide range of career opportunities. This flexibility stems from the diversity of skills acquired and the sustained demand for marketing professionals in both traditional and digital industries.
- Internship experience and employability ($r = 0.774$, $p \leq 0.001$). Internships significantly enhance employability by providing students with applied experience, which is highly valued by employers (Cranmer, 2006). These findings underscore the importance of university-industry collaboration in facilitating the transition from education to employment.
- Critical thinking and digital competencies ($r = 0.770$, $p \leq 0.001$). In a digital economy, critical thinking and digital literacy are complementary. The ability to engage in deep analysis and to adapt to technological changes is essential for professional success in increasingly digitalized environments (Van Laar et al., 2017).
- Previous work experience and recruitment outcomes ($r = 0.753$, $p \leq 0.001$). Graduates with prior work experience—whether through part-time jobs, volunteering, or internships—are perceived as better prepared for the professional environment, which increases their chances of being selected in recruitment processes.
- University reputation and academic performance ($r = 0.750$, $p \leq 0.001$). The prestige of the university positively influences employer perceptions, opening access to more competitive and better-paid positions (Brown & Hesketh, 2004). This highlights the symbolic capital conferred by the institution of graduation.
- Legal studies and employability ($r = 0.750$, $p \leq 0.001$). Graduates with legal specializations—particularly in areas such as digital law or compliance in technological fields—demonstrate high employability, reflecting the growing need for legal regulation in emerging technological domains.
- Social skills and communication ($r = 0.745$, $p \leq 0.001$). Interpersonal competencies are critical for teamwork and workplace integration. Effective communication is a key requirement across multiple professional sectors, and well-developed social skills enhance adaptability in organizational settings.
- Engineering fields and job stability ($r = 0.713$, $p \leq 0.001$). Engineering remains a domain with consistent demand, particularly in emerging industries such as

renewable energy and cybersecurity. The correlation confirms long-term career stability and high employability in this sector.

- IT sector and labor market demand ($r = 0.660$, $p \leq 0.001$). Advanced digital skills and IT specializations are among the most sought-after qualifications in the labor market, underlining the critical role of technological competencies in the contemporary economy (OECD, 2021).

The analysis reveals that graduate integration into the labor market is strongly influenced by a combination of academic specialization, practical experience, and transversal competencies. Fields such as Marketing, Law, Engineering, and IT demonstrate high adaptability and demand, while soft skills, critical thinking, and digital literacy significantly enhance employability. These findings underscore the importance of aligning higher education with labor market requirements and promoting experiential learning opportunities to foster graduates' long-term professional success.

Using Principal Component Analysis at the Level of Research Variables

Principal Component Analysis (PCA) is a fundamental statistical technique widely employed for dimensionality reduction, with the primary aim of preserving as much of the original information as possible. This method transforms a set of correlated variables into a new set of uncorrelated variables known as principal components. These components are linear combinations of the original variables and are constructed in such a way as to capture the maximum variance in the data (Field, 2018). PCA is essential in academic research due to its ability to facilitate data interpretation, eliminate redundancy, and improve the predictive performance of models across various disciplines, including economics, psychology, social sciences, and labor market analysis.

Mathematically, PCA relies on eigenvalue decomposition of the covariance or correlation matrix, allowing the identification of directions (components) that represent the greatest variance in the dataset (Tabachnick & Fidell, 2019). The PCA procedure involves several key steps: standardization of data to ensure comparability across variables, computation of the covariance matrix to examine inter-variable relationships, extraction of eigenvectors and eigenvalues to determine the principal components, selection of the most informative components based on explained variance, and transformation of the original dataset into the newly defined, lower-dimensional space (Jolliffe & Cadima, 2016).

The application of PCA offers multiple advantages in empirical research. It reduces the dimensionality of large datasets, thereby simplifying analysis while preserving essential information (Ringnér, 2008). PCA also helps identify latent structures within the data, highlighting the underlying relationships among variables and enhancing interpretability (Hair et al., 2019). Furthermore, it improves data visualization by projecting high-dimensional data onto two- or three-dimensional plots, enabling clearer insights into trends and groupings. A significant methodological benefit of PCA lies in its ability to prevent multicollinearity, which strengthens the reliability and accuracy of predictive models (Tabachnick & Fidell, 2019).

In the context of research on the determinants of graduate employability, PCA serves as a valuable analytical tool for identifying the most influential variables that contribute to professional success. By applying PCA, researchers can reduce the number of variables under consideration and isolate the principal components that explain most of the variance in employability outcomes. This approach enhances data interpretation and supports the formulation of evidence-based conclusions relevant to both the educational and economic domains (Ringnér, 2008) (Table 4.1).

The results of the Kaiser-Meyer-Olkin (KMO) test and Bartlett's Test of Sphericity indicate that the dataset is well-suited for factor analysis. The KMO value of 0.81 suggests a high level of sampling adequacy, implying that the variables are sufficiently intercorrelated to allow for the extraction of meaningful latent factors. A KMO value above 0.80 is generally considered "very good" and reflects a solid foundation for conducting PCA (Field, 2018). This means that the dataset possesses the internal structure necessary for effective dimensionality reduction.

Furthermore, Bartlett's Test of Sphericity yields a highly significant p-value ($p < 0.001$), which confirms that the correlation matrix is not an identity matrix. This result indicates the presence of statistically significant relationships among the variables, validating the appropriateness of PCA as a method for uncovering latent structures within the data (Hair et al., 2019). The significance of this test reinforces the idea that the variables are sufficiently correlated to justify their reduction into a smaller set of unobserved components (Fig. 4.1).

The Scree Plot in Principal Component Analysis (PCA) is a crucial tool for determining the optimal number of factors to retain in the final model. This plot graphically represents the variation in eigenvalues across the extracted components, highlighting the point at which the contribution of each component to explaining the variance begins to level off. According to the elbow rule, the inflection point of the curve indicates the optimal number of components to retain, as subsequent components add minimal additional information to the model (Jolliffe, 2002).

In interpreting the scree plot, the focus is on identifying the point of inflection, where the eigenvalue curve changes steeply, followed by a gradual leveling off. This method is supported by Kaiser's Rule, which suggests retaining components with eigenvalues greater than 1 (Kaiser, 1960). When the first few components explain

Table 4.1 KMO and Bartletts' Test

Kaiser-Meyer-Olkin Measure of Sampling Adequacy.		0.81
Bartlett's Test of Sphericity	Approx. Chi-Square	8158.355
	df	2628
	Sig.	0.000

Source: Compiled by the author using SPSS based on the existing database

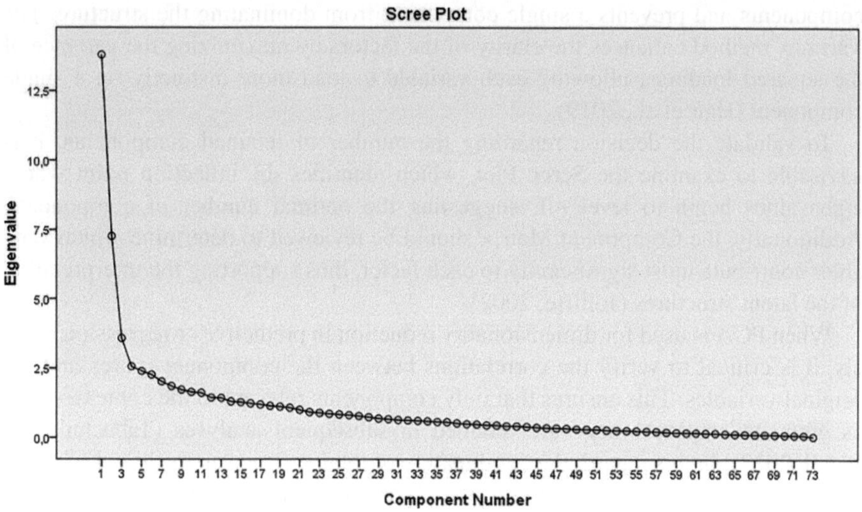

Fig. 4.1 Scree plot PCA. (Source: Compiled by the author using SPSS based on the existing database)

more than 70–80% of the total variance, dimensionality reduction is considered efficient and justifiable (Tabachnick & Fidell, 2013).

From a methodological standpoint, the choice of the number of components to retain should be cross-checked with the analysis of the Total Variance Explained table to confirm the decision regarding the number of components selected. In this context, interpreting the scree plot should be complemented by an examination of the component matrix to identify which variables contribute most to each factor, thereby facilitating a clearer interpretation of the latent dimensions identified through PCA (Hair et al., 2019).

The Total Variance Explained table in Principal Component Analysis (PCA) offers a detailed overview of how well the extracted components represent the initial variables. Eigenvalues reflect the amount of variance accounted for by each component, and the choice of extraction and rotation methods significantly influences the interpretability of the resulting factors (Kaiser, 1960; Jolliffe, 2002).

In the current analysis, the first component has an eigenvalue of 13.794, explaining 18.89% of the total variance. This indicates a substantial influence on the overall data structure. According to Kaiser's criterion, components with eigenvalues equal to or greater than 1 should be retained. Based on this rule, the first 20 components are kept, collectively accounting for 70.94% of the total variance—demonstrating an effective and justified dimensionality reduction (Kaiser, 1960; Tabachnick & Fidell, 2013).

The application of Varimax rotation redistributes the explained variance more evenly across components to enhance interpretability. Following rotation, the first component explains 9.36% of the total variance, down from 18.89% prior to rotation. This redistribution reflects a more balanced contribution of variables to the components and prevents a single component from dominating the structure. The Varimax method enhances the clarity of the factors by maximizing the variance of the squared loadings, allowing each variable to load more distinctly on a single component (Hair et al., 2019).

To validate the decision regarding the number of retained components, it is advisable to examine the Scree Plot, which identifies the inflection point where eigenvalues begin to level off, suggesting the optimal number of components. Additionally, the Component Matrix should be reviewed to determine which variables contribute most significantly to each factor, thus supporting the interpretation of the latent structures (Jolliffe, 2002).

When PCA is used for dimensionality reduction in predictive or regression models, it is critical to verify the correlations between the component scores and the original variables. This ensures that only components relevant to the context—such as graduate employability—are retained in subsequent analyses (Tabachnick & Fidell, 2013). In this regard, the combination of total variance analysis and Varimax rotation contributes meaningfully to the clarification of the data structure and the enhancement of interpretability.

Moreover, the interpretability of the extracted components is essential for establishing a meaningful connection between statistical results and theoretical constructs related to graduate employability. Interpreting factor loadings allows researchers to identify latent dimensions that correspond to core aspects of higher education outcomes, such as cognitive competencies, transferable skills, and institutional engagement (Yorke, 2006; Bridgstock, 2009). This process ensures that the statistical reduction of data does not compromise conceptual clarity but rather facilitates the translation of empirical patterns into theoretically grounded insights. As noted by Jolliffe and Cadima (2016), the value of principal component analysis lies not only in its ability to reduce redundancy but also in its potential to reveal underlying structures that support policy-relevant interpretations in education and labor market research.

The findings derived from the correlation analysis and principal component analysis (PCA) offer a multidimensional perspective on the mechanisms that underpin graduate students' employability in the context of a knowledge-driven economy. By identifying statistically significant relationships between key educational variables and labor market outcomes, the study provides robust empirical evidence on the relevance of academic specialization, practical experience, transversal skills, and institutional prestige in shaping professional trajectories.

The correlation matrix has revealed strong associations—statistically significant at the 0.001 level—between variables such as internship experience, digital and critical thinking skills, prior employment exposure, and successful recruitment outcomes. These results underscore the growing emphasis placed by employers on applied competencies and behavioral adaptability, confirming existing theoretical assumptions in human capital and employability literature. Notably, fields such as Marketing, Engineering, IT, and Law exhibit high levels of labor market adaptability, suggesting that specialization aligned with market demand yields greater employment stability and flexibility.

The application of Principal Component Analysis substantially enhances the analytical rigor of the study by allowing the identification of latent constructs that explain the observed variation in employability-related outcomes. The data demonstrated strong suitability for factor analysis, as evidenced by a Kaiser-Meyer-Olkin measure of 0.81, indicative of meritorious sampling adequacy, and a highly significant result from Bartlett's Test of Sphericity, which confirms the structural validity of the variable set under consideration. The first 20 extracted components accounted for approximately 71% of the total variance, demonstrating a high level of informational efficiency within the reduced data structure, as detailed in Appendix 1.

The integration of PCA into the analytical framework facilitates the identification of core constructs—such as practical engagement, institutional capital, and soft skill proficiency—that underpin successful labor market integration. This dimensionality reduction not only refines the interpretation of complex data structures but also informs evidence-based educational policy and curriculum reform. By isolating the most influential factors, PCA supports a strategic reorientation of higher education programs towards competencies that are demonstrably aligned with employer expectations.

In sum, the dual application of correlation analysis and PCA demonstrates the critical role of empirical methods in uncovering the systemic determinants of graduate employability. The results substantiate the imperative for higher education institutions to embed experiential learning, digital literacy, and transferable skill development into academic programs. Simultaneously, they highlight the value of advanced statistical modeling in producing actionable insights for stakeholders invested in the optimization of educational outcomes and workforce integration.

Perspective of Employers and the Assessment of Competence of Graduates

In this part of the paper, we investigated the perspective of employers on skills assessment in the labor market. We assessed the skills sought, the prioritization of candidates, and the advice offered to graduates, examining the current and future occupational landscape.

The sample analyzed offers a comprehensive overview of the labor market's diversity, encompassing companies from various sectors and regions. In this context, SC Versus SRL from Suceava, operating in the real estate industry since 2003 with a staff of five, reflects the structure and scale of small local enterprises. Conversely, OVB Allfinanz Romania, active in financial intermediation since 2002 with approximately 5500 employees nationwide, illustrates the presence of large-scale organizations in the Romanian economy. Similarly, Raiffeisen Bank in Suceava, founded in 1998 and employing 22 individuals, highlights the significance of the banking sector in the region.

The sample also includes companies such as Transilvania Gold Events, which has been operating in the events industry since 2010 with 10 employees, showcasing the expansion of service-oriented businesses. In the field of information technology, Expertware SRL, an IT company based in Suceava with 90 employees and established in 2006, offers insight into the regional tech sector. Furthermore, SC Nordic Residence SRL, a rental company launched in 2023 in Suceava, and SC Myrotrip Flight SRL from Botoşani, active in tourism since 2014, demonstrate growing interest in the development of the local tourism industry.

Additionally, SC Dormarioplus SRL, a retail company from Vicovu de Sus with four employees since 2008, and Patachi Express, a transportation firm established in 2020 in the same locality with four employees, illustrate the diversity and role of small businesses in the local economy. In the field of research and development, Exact Business Solutions, with 30 employees and operations in Bucharest, Suceava, and Botoşani since 2013, exemplifies the regional engagement in knowledge-based sectors. Finally, SC Office Expert SRL, an accounting services firm founded in 2008 in Suceava with five employees, highlights the importance of professional services in supporting the local economic framework.

This diverse array of companies contributes to a multidimensional perspective on the regional labor market, emphasizing each firm's role in shaping both local and national economic dynamics.

Digital Skills

The most sought-after skills by employers are interpersonal skills (34.8%), followed by domain-specific technical skills (26.1%), digital competencies (17.4%), time-constrained problem-solving abilities (13%), and leadership skills (8.7%). In this regard, Fig. 4.2 summarizes the priorities expressed by employers when evaluating graduates' applications.

According to employers, the main methods for acquiring professional experience include internships (mentioned by 7 employers), developing technical skills outside the university curriculum, participating in mentorship programs (4 employers), and engaging in volunteer activities to build networking opportunities. In this context, Fig. 4.3 illustrates the recommendations provided to young graduates regarding their integration into the labor market. Employers' portfolio of achievements.

Figure 4.4 illustrates the employers' recommendations for graduates, highlighting the key aspects considered essential for their success in the labor market. Each category represents a valuable skill or practice, while the scores indicate the relative importance assigned by employers.

The most critical criterion, with a score of 8, is participation in internships, highlighting how highly employers value practical experience. This is followed by

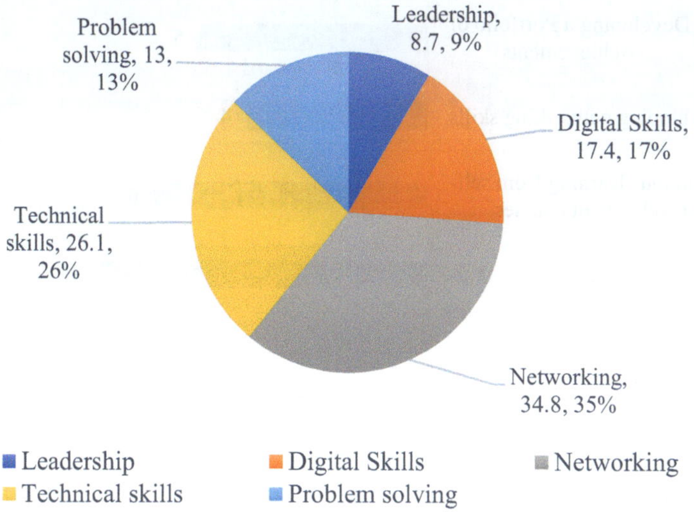

Fig. 4.2 Skills wanted by employers. (Source: Compiled by the author using SPSS based on the existing database)

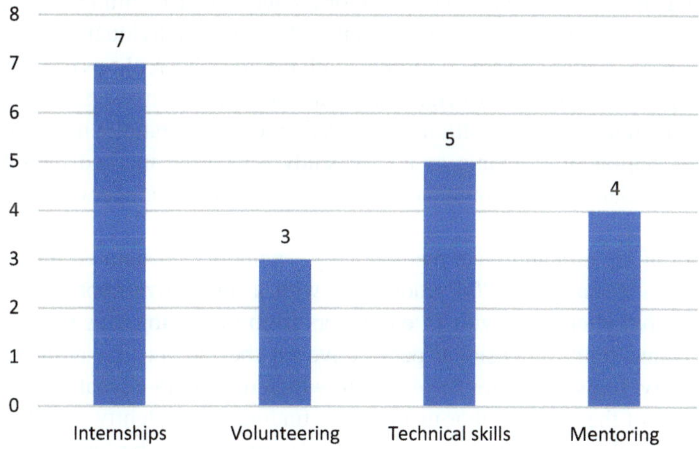

Fig. 4.3 Ways for students to gain experience. (Source: Compiled by the author using SPSS based on the existing database)

continuous learning through personal development courses, scoring 6, which emphasizes the need for ongoing adaptation and skill enhancement. Developing a portfolio of achievements received a score of 5, indicating the importance of showcasing results and competencies. Lastly, networking skills, with a score of 4, reflect the relevance of professional relationships, although they are considered less of a priority compared to other criteria.

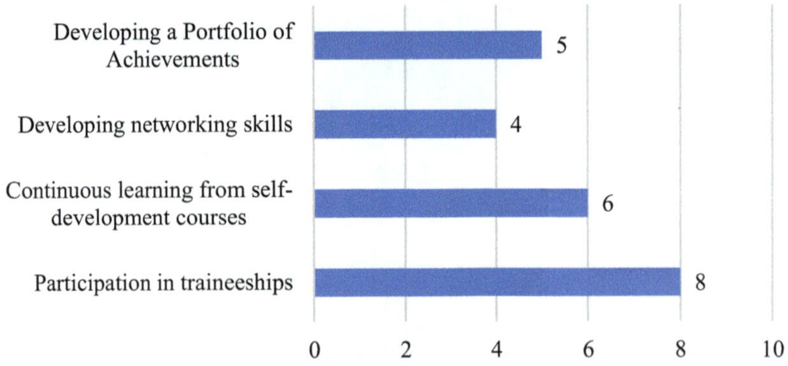

Fig. 4.4 Employers' recommendations for graduates. (Source: Compiled by the author using SPSS based on the existing database)

4.5 Conclusion

The integration of higher education graduates into the labor market represents one of the fundamental challenges of contemporary society—an increasingly relevant topic that generates constant concern among educational institutions, employers, policymakers, and, perhaps most importantly, the young individuals who complete their university studies and are confronted with the complex realities of professional life. This paper has sought to rigorously and objectively investigate how higher education facilitates the transition from academic life to employment, using the North-East region of Romania as a case study—a region marked by significant structural imbalances, yet simultaneously rich in potential for economic and social development.

By combining a quantitative methodology—namely, a structured questionnaire distributed to a sample of 200 graduates—with a qualitative approach based on semi-structured interviews with 11 employers from the region, this research offers a nuanced and multidimensional perspective on the factors that influence youth employability. It also explores the extent to which university education aligns with the demands of the current labor market. The findings consistently reveal that post-graduation employment rates are not solely determined by the level of education attained but are even more strongly influenced by the nature of the competencies acquired, the relevance of the field of study to current economic demand, and the presence of practical work experience accumulated during university studies.

The validation of the research objectives, set at the beginning of this study, emerges naturally from the empirical data. The relationship between the field of study and successful labor market insertion is clear, though nuanced. Degrees in fields such as information technology, engineering, marketing, and law were found to be the most effective in supporting a rapid transition into the workforce. Conversely, graduates in the humanities and social sciences often find themselves working in fields not aligned with their academic specialization. This trend underscores the importance of aligning academic programs with economic needs and of providing career guidance early on in students' academic journeys.

The impact of acquired competencies on professional success was also empirically demonstrated. Graduates with strong digital skills, who had undertaken internships or part-time jobs, and who had developed soft skills such as communication, adaptability, and teamwork, experienced smoother and more efficient labor market integration. These findings are consistent with broader European trends and confirm that success in today's job market is increasingly dependent on a combination of technical expertise and interpersonal competencies.

The importance of strategic partnerships between universities and the economic sector was highlighted by both groups of respondents—graduates and employers alike. Employers repeatedly emphasized that practical experience and institutional collaboration were key differentiators during the candidate selection process. From internship opportunities to collaborative research projects and guest lectures from industry professionals, these partnerships serve as bridges between theoretical knowledge and real-world application.

Graduates who participated in such partnerships or who benefited from institutional support in identifying employment opportunities were generally more successful in securing jobs aligned with their academic preparation. In addition, individual job-search strategies played a crucial role. Graduates who built personal portfolios, who engaged in professional networking, and who had access to career counseling services not only found employment more quickly but also were more likely to work in their field of study.

One of the most significant challenges identified in the research is the persistent mismatch between the skills developed during university education and those demanded by employers. This gap was clearly articulated by employers during the interviews and was supported by the quantitative data. While most employers acknowledge that recent graduates possess a solid theoretical foundation, they also note a lack of practical skills, limited problem-solving abilities under pressure, insufficient ownership of tasks, and a generally passive attitude that fails to meet the proactive standards of modern workplaces.

These findings point to the urgent need for a comprehensive curricular reform in higher education. Universities must reposition themselves as dynamic institutions capable of anticipating labor market changes and preparing students accordingly. This means moving away from strictly content-focused education toward more flexible, skills-based learning that incorporates real-life problem-solving, project-based assignments, and applied research components.

Experiential learning—in the form of internships, extracurricular projects, or volunteer work—emerged as the single most influential factor in determining professional success. The data collected indicates a statistically significant correlation between such forms of experiential learning and the employment rate among graduates. This strongly suggests that applied learning must become a central component of the university curriculum. It is not enough for students to accumulate academic knowledge; they must also learn to translate that knowledge into practical action and demonstrate it in real-world contexts.

Furthermore, soft skills such as effective communication, emotional intelligence, and critical thinking were consistently valued across all employment sectors. These transversal competencies are no longer "nice to have" attributes; they are indispensable for long-term career success. Thus, institutions of higher education must broaden their educational scope and equip students with the tools they need to navigate increasingly complex and interdisciplinary work environments.

In light of these findings, it becomes evident that universities must adopt a more proactive role in shaping not only the intellectual capacity of their graduates but also their professional readiness. This involves rethinking traditional educational models and focusing more intensively on interdisciplinary approaches, entrepreneurship education, and real-world problem-solving. Strong and sustainable university-industry collaborations, the integration of work-based learning into degree programs, and the creation of well-functioning career development centers are essential for facilitating students' transition from academia to the labor market.

Additionally, universities must strive to enhance their institutional reputation—not only through academic achievements but also through their capacity to produce graduates who meet and exceed labor market expectations. By establishing themselves as trusted partners to employers and contributing actively to regional innovation ecosystems, universities can solidify their role as engines of sustainable development.

The North-East region of Romania, despite its multiple structural challenges, has considerable potential to become a center of economic and social revitalization. Its educational institutions can play a key role in this transformation by contributing to the retention of the local youth workforce, by attracting investment and by stimulating innovation. For this potential to be realized, however, higher education programs must be closely aligned with current and anticipated labor market needs. This means continuously updating curricula, forming advisory boards with industry representatives, and leveraging digital technologies to modernize teaching and learning processes.

Moreover, public policy must support this transformation. Targeted funding for university business partnerships, incentives for companies that provide internships, and national strategies for the development of digital and transversal competencies are all essential components of a coherent, forward-looking educational ecosystem.

In conclusion, higher education must redefine its mission and take on a central role in the socioeconomic ecosystem. It must contribute actively to the development of human capital, reduce youth unemployment, and support the sustainable development of regions across the country. This can only be achieved through an integrated vision that prioritizes collaboration, innovation, and responsiveness to the real needs of society.

Universities should not simply provide degrees; they should offer pathways to meaningful, fulfilling, and sustainable careers. Only by embracing this vision can higher education institutions rise to the challenges of the twenty-first century and provide their graduates with more than academic credentials—they must offer the knowledge, skills, and attitudes necessary to thrive in an increasingly dynamic and demanding world.

Appendix

Appendix 1. The Total Variance Explained Table in Principal Component Analysis (PCA)

	Initial Eigenvalues			Extraction Sums of Squared Loadings			Rotation Sums of Squared Loadings		
Component	Total	% of Variance	Cumulative %	Total	% of Variance	Cumulative %	Total	% of Variance	Cumulative %
1	13.794	18.896	18.896	13.794	18.896	18.896	6.834	9.362	9.362
2	7.254	9.938	28.834	7.254	9.938	28.834	4.694	6.43	15.792
3	3.578	4.91	33.735	3.578	4.91	33.735	4.351	5.96	21.753
4	2.56	3.57	37.242	2.56	3.57	37.242	4.348	5.957	27.79
5	2.391	3.276	40.518	2.391	3.276	40.518	3.688	5.052	32.761
6	2.256	3.09	43.67	2.256	3.09	43.67	3.232	4.427	37.189
7	2.013	2.758	46.365	2.013	2.758	46.365	2.69	3.574	40.762
8	1.848	2.531	48.896	1.848	2.531	48.896	2.61	3.564	44.326
9	1.716	2.351	51.247	1.716	2.351	51.247	2.484	3.43	47.729
1	1.637	2.242	53.49	1.637	2.242	53.49	1.828	2.54	50.233
11	1.65	2.199	55.688	1.65	2.199	55.688	1.746	2.391	52.624
12	1.432	1.962	57.651	1.432	1.962	57.651	1.634	2.238	54.862
13	1.423	1.949	59.599	1.423	1.949	59.599	1.577	2.161	57.023
14	1.296	1.775	61.374	1.296	1.775	61.374	1.559	2.136	59.158
15	1.269	1.738	63.112	1.269	1.738	63.112	1.553	2.127	61.286
16	1.223	1.675	64.788	1.223	1.675	64.788	1.478	2.024	63.31
17	1.184	1.622	66.41	1.184	1.622	66.41	1.422	1.947	65.258
18	1.133	1.552	67.962	1.133	1.552	67.962	1.46	1.926	67.184
19	1.19	1.519	69.481	1.19	1.519	69.481	1.386	1.899	69.083
20	1.068	1.464	70.944	1.068	1.464	70.944	1.359	1.862	70.944
21	0.992	1.359	72.34						
22	0.91	1.247	73.551						
23	0.883	1.21	74.761						
24	0.849	1.162	75.923						
25	0.828	1.134	77.057						
26	0.86	1.14	78.161						
27	0.775	1.061	79.222						
28	0.732	1.003	80.226						
29	0.685	0.939	81.164						
3	0.663	0.98	82.072						
31	0.634	0.868	82.94						
32	0.618	0.847	83.787						
33	0.63	0.827	84.614						
34	0.59	0.88	85.422						

Component	Initial Eigenvalues			Extraction Sums of Squared Loadings			Rotation Sums of Squared Loadings		
	Total	% of Variance	Cumulative %	Total	% of Variance	Cumulative %	Total	% of Variance	Cumulative %
35	0.567	0.776	86.198						
36	0.56	0.767	86.964						
37	0.519	0.711	87.675						
38	0.489	0.669	88.344						
39	0.479	0.657	89.001						
4	0.448	0.614	89.616						
41	0.444	0.68	90.224						
42	0.418	0.573	90.797						
43	0.413	0.566	91.363						
44	0.395	0.541	91.94						
45	0.369	0.55	92.49						
46	0.353	0.483	92.892						
47	0.346	0.473	93.366						
48	0.332	0.455	93.821						
49	0.38	0.423	94.244						
5	0.31	0.413	94.657						
51	0.294	0.43	95.06						
52	0.273	0.373	95.433						
53	0.259	0.354	95.788						
54	0.255	0.35	96.137						
55	0.245	0.335	96.472						
56	0.23	0.315	96.787						
57	0.223	0.35	97.093						
58	0.198	0.271	97.364						
59	0.188	0.258	97.621						
6	0.181	0.248	97.87						
61	0.178	0.244	98.114						
62	0.166	0.227	98.341						
63	0.164	0.225	98.567						
64	0.139	0.19	98.757						
65	0.131	0.18	98.937						
66	0.121	0.166	99.13						
67	0.119	0.162	99.265						
68	0.116	0.159	99.425						
69	0.18	0.148	99.573						
7	0.097	0.133	99.76						
71	0.092	0.126	99.832						
72	0.081	0.112	99.944						
73	0.041	0.056	100.000						

Extraction Method: Principal Component Analysis

References

Arionesei, G., Hapenciuc, C.-V., & Costea, M. (2016). Statistical confrontation of the evolution of tourism in the north-east region in comparison with the other regions of Romania. *Amfiteatru Economic, 18*(10), 798–814.

Becker, G. S. (1993). *Human capital: A theoretical and empirical analysis, with special reference to education* (3rd ed.). University of Chicago Press.

Brennan, J. (2010). *Higher education and its communities: Interconnections and interdependencies*. Society for Research into Higher Education, 56, 303–324.

Bridgstock, R. (2009). The graduate attributes we've overlooked: Enhancing graduate employability through career management skills. *Higher Education Research & Development, 28*(1), 31–44.

Brown, P., & Hesketh, A. (2004). *The mismanagement of talent: Employability and jobs in the knowledge economy*. Oxford University Press.

Bryman, A., & Cramer, D. (2020). *Quantitative data analysis with SPSS for windows*. Routledge.

Buta, S., Kicsi, R., & Cozorici, N. (2010). Empirical reflections regarding the benefits of human capital. The annals of the "Ştefan cel Mare" University of Suceava. *Fascicle of the Faculty of Economics and Public Administration, 10*(1), 131–145.

Cohen, J., Cohen, P., West, S. G., & Aiken, L. S. (2013). *Applied multiple regression/correlation analysis for the behavioral sciences* (3rd ed.). Routledge.

Cortese, A. D. (2003). The critical role of higher education in creating a sustainable future. *Planning for Higher Education, 31*(3), 15–22.

Cranmer, S. (2006). Enhancing graduate employability: Best intentions and mixed outcomes. *Studies in Higher Education, 31*(2), 175–190.

de Assis Dornelles, J. A., Ayala, N. F., & Frank, A. G. (2022). Smart working in Industry 4.0: How digital technologies enhance manufacturing workers' activities. *Computers & Industrial Engineering, 163*, 28–35.

Dimian, G. (2014). Labor market obstacles faced by young professionals. *Journal of Social and Economic Statistics, 3*(2), 45–56.

DiPrete, T. A., & Nonnemaker, L. (1997). Structure and dynamics of the labor market: Implications for social equity. *Social Science Research, 26*(3), 150–201.

Eldridge, D. (2011). Labor market dynamics: Core interactions and effects. *Economics and Social Policy, 29*(4), 15–34.

Field, A. (2018). *Discovering statistics using IBM SPSS statistics* (5th ed.). Sage.

Gaube, G. (2015). The impact of tourism on society and the environment. *Tourism Journal, Bucharest University Press, 12*(3), 45–67.

Hair, J. F., Black, W. C., Babin, B. J., & Anderson, R. E. (2019). *Multivariate data analysis* (8th ed.). Cengage Learning.

Harvey, L. (2003). Transitions from higher education to work. *European Journal of Education, 38*(4), 425–446.

Heckman, J. (1979). Sample selection bias as a specification error. *Econometrica, 47*(1), 153–161.

Institutul Naţional de Statistică. (2023). *Forţa de muncă în România – ocupare şi şomaj*. INS.

Jabłoński, A. (2021). Human capital theory reconsidered: Strengths and weaknesses. *Journal of Human Capital Studies, 15*(2), 145–158.

Janice, T., Wilson, M., & Carter, E. (2016). Human capital: Economic perspectives. *Journal of Human Capital Studies, 10*(4), 131–145.

Jolliffe, I. T. (2002). *Principal component analysis*. Springer.

Jolliffe, I. T., & Cadima, J. (2016). Principal component analysis: A review and recent developments. *Philosophical Transactions of the Royal Society a Mathematical, Physical and Engineering Sciences, 374*(2065), 20150202.

Kaiser, H. F. (1960). The application of electronic computers to factor analysis. *Educational and Psychological Measurement, 20*(1), 141–151.

Mascaradi, L., Reilly, F., & Turner, J. (2022). Educational innovation and workforce development: Bridging theory and practice. *Journal of Human Capital Studies, 6*, 100–115.
Mincer, J. (1958). *Investment in human capital and personal income distribution* (pp. 281–302). Princeton University Press.
Mincer, J. (1974). *Schooling, experience, and earnings*. Columbia University Press.
Neamtu, G., & Burac, D. (2023). Organizational performance and human capital in emerging economies. *Journal of Economic Development and Organizational Studies, 15*(2), 22–34.
OECD. (2021). *Skills for a digital world.* [Online] https://www.oecd.org/skills/. Accessed 15 Mar 2025.
Patache, L. (2011). Higher education and the labor market – Romania case of study. *Journal of Academic Research in Economics, XI*(1), 1614–1619.
Reder, M. W., & Becker, G. (1967). Gary Beckers human capital: A review article. *The Journal of Human Resources, 2*, 97.
Ringnér, M. (2008). What is principal component analysis? *Nature Biotechnology, 26*(3), 303–305.
Schultz, T. (1960). Capital formation by education. *Journal of Political Economy, 68*(6), 571–583.
Schultz, T. W. (1968). Resources for higher education: An economist's view. *Journal of Political Economy, 76*(3), 327–347.
Șușu, Ș. (2007). Rolul turismului în regiunea de dezvoltare Nord-Est. *Revista de Turism, 3*, 63–66.
Tabachnick, B. G., & Fidell, L. S. (2013). *Using multivariate statistics*. Pearson.
Tabachnick, B. G., & Fidell, L. S. (2019). *Using multivariate statistics* (7th ed.). Pearson.
Teiceira, P. (2014). The new economics of human capital: Theory and applications. *IZA Journal of Labor Economics, 3*(1), 93–101.
Tomlinson, M. (2017). Forms of graduate capital and their relationship to graduate employability. *Education + Training, 59*(4), 338–352.
Ugwu, C., & Eze, V. (2023). Qualitative research. *IDOSR Journal of Computer and Applied Sciences, 8*(1), 20–35.
van Laar, E., van Deursen, A. J. A. M., van Dijk, J. A. G. M., & de Haan, J. (2017). The relation between 21st-century skills and digital skills: A systematic literature review. *Computers in Human Behavior, 72*, 577–588.
Yorke, M. (2006). *Employability in higher education: What it is – What it is not. Learning and employability series 1* (pp. 1–18). Higher Education Academy.

Chapter 5
Adapting Vocational Education for Industry 5.0 by Transforming the Professional Identity of VET Trainers in Wood Processing

Horia-Dominic Schreiner, Bogdan Mocan, and Mircea Fulea

Abstract This study develops a theoretical framework for transforming vocational education and training (VET) for the emerging Industry 5.0 paradigm by focusing on the critical role of VET trainers' professional identity. As Industry 5.0 emphasizes human-machine collaboration, artificial intelligence integration, and sustainability, traditional VET programs—particularly in craft-based sectors like wood processing—require fundamental reconceptualization. While research has addressed curriculum reforms and technological adoption, limited attention has been given to how VET trainers must transform their professional identity and practices to effectively develop an agile workforce. We propose a conceptual framework for integrating Industry 5.0 approaches into VET programs, specifically applied to a wood processing technician education program in the Romanian VET system, and identify the key competencies for an agile workforce in the Industry 5.0 context. It further proposes a redesign of industry-partnered training modules and identifies how VET trainers must evolve from traditional instructors to facilitators of human-machine collaboration, developing competencies in digital tools, adaptive pedagogy, and sustainability practices. While requiring empirical validation, the research contributes a structured approach for aligning vocational training with Industry 5.0 requirements, offering a roadmap for educators and industry stakeholders to update VET programs for future labor market demands. It also provides the groundwork for targeted professional development strategies to equip VET trainers with the necessary skills to prepare an agile, technology-enabled, and sustainability-conscious workforce.

Keywords Vocational education · Industry 5.0 · Wood processing · Workforce agility

H.-D. Schreiner · B. Mocan · M. Fulea (✉)
Technical University of Cluj-Napoca, Cluj-Napoca, Romania
e-mail: bogdan.mocan@muri.utcluj.ro; mircea.fulea@staff.utcluj.ro

A. L. Negrusa, M. M. Coroş (eds.), *Innovation, Sustainability, and Growth in a VUCA Environment*, Springer Proceedings in Business and Economics,
https://doi.org/10.1007/978-3-032-11639-0_5

5.1 Introduction

The foreseeable transition to Industry 5.0 presents both challenges and opportunities for vocational education and training (VET) programs. Industry 5.0 emphasizes human-machine collaboration, artificial intelligence integration, and sustainability, requiring a fundamental shift in workforce skills and training methodologies (Nahavandi, 2019; Xu et al., 2021). This study explores how vocational education can support the development of an agile workforce in the context of Industry 5.0 by transforming the professional identity of VET trainers (Pilz & Wiemann, 2021).

The research problem addressed in this study is: How can vocational education support the development of an agile workforce in the context of Industry 5.0 through the transformation of VET trainers' professional identity? The study focuses on VET in wood processing programs. Here, traditional training approaches, often focused on manual skills and craftsmanship, must now incorporate digital tools, automation, and sustainable practices. This shift requires not only curriculum adjustments but also a redefinition of the role of VET trainers as facilitators of human-machine collaboration (Marope et al., 2015; Alam & Mohanty, 2023).

While the transition to Industry 5.0 has raised growing interest in updating vocational education to include digital tools, automation, and sustainability, limited research can be found on how this transition impacts the professional identity of VET trainers—particularly in traditional, craft-based sectors such as wood processing (Cedefop, 2020; Chankseliani & Anuar, 2019). While the existing literature focuses on curriculum reform, technology adoption, or learner skill development, the key role of trainers as agents of change seems less discussed. There is a need for a deeper understanding of how trainers perceive and adapt their roles, teaching practices, and values in response to Industry 5.0, and how this transformation can enable the formation of an agile, human-centered workforce (Susskind, 2015; Autor, 2015).

This research gap is important because successful Industry 5.0 integration depends not only on updated curricula but on trainers who can effectively bridge traditional craftsmanship with digital competencies. Without understanding how trainers' professional identity should evolve, VET programs risk implementing superficial technological changes that fail to develop truly agile workers.

This study addresses this gap by developing a conceptual framework for integrating Industry 5.0 approaches into VET programs, in particular for an upper secondary education program training wood processing technicians. Based on the existing training modules and the competencies defined by the occupational standard, we propose a set of "custom" training modules with relevance for Industry 5.0. We then identify the competencies required by trainers to successfully deliver these modules.

The rest of the document is structured as follows. Section 5.2 discusses the theoretical background and introduces the conceptual framework of the study. Section 5.3 describes the actual steps taken, from the existing training modules and the competencies defined by the occupational standard to the (new) competencies of the VET trainers coming from the industrial partners. Section 5.4 discusses the results and the contributions.

5.2 Background

This study develops a conceptual framework for integrating Industry 5.0 approaches into VET programs (Maddikunta et al., 2022). Although the framework is designed to be general, we focus in this research on a particular VET program—wood processing technician (vocational upper secondary education program, grades 9–12), one of the authors being involved in such a program.

The framework is built upon three core dimensions. The first dimension covers integration of advanced technologies that can (and must) be incorporated into the curricula, such as CAD/CAM software for digital design, CNC machines for automated wood processing, and collaborative robotics for precision manufacturing, productivity, and superior quality (Thoben et al., 2017).

The second dimension implies identification of the most important competencies of the professional standard for an agile workforce: as Industry 5.0 emphasizes adaptability and cross-disciplinary expertise (Demir et al., 2019), wood processing technicians must develop new technical skills beyond traditional woodworking. They should build digital competencies (e.g., proficiency in CAD/CAM design, CNC programming, and robotic automation), sustainability awareness (e.g., understanding resource-efficient production, using renewable materials, and minimizing environmental impact), and soft skills (collaboration, adaptability, and critical thinking) (McGrath et al., 2020).

The third dimension focuses on (re)designing particular training modules of the VET curriculum. For each year of study, the curriculum foresees one training module to be designed and run by local industrial organizations that partner in the VET program. For these particular modules, we propose an Industry 5.0-aligned curriculum that integrates traditional skills (core wood processing techniques such as cutting, joining, and finishing), technological advancements (use of digital tools and automation for precision and efficiency), and soft skills and adaptability (preparing learners for a rapidly evolving industry by fostering problem-solving and teamwork skills) (Schröder, 2019; Hirschi, 2018).

The conceptual framework is illustrated in Fig. 5.1.

The first layer, corresponding to the external context that demands transformation, depicts the Driving Forces of Industry 5.0: Human-Machine Collaboration & AI, Digitalization & Automation, Sustainability Goals, and Changing Labor Market Needs.

The second layer, the core of our framework, shows the proposed VET system transformation. Curriculum innovation (what is taught), grounded in the two above dimensions, integrates Traditional Skills + CAD/CAM + CNC, Additive manufacturing (e.g., 3D wood printing), and Soft skills: adaptability, teamwork, critical thinking. The Trainer Role Evolution (How it is taught) implies transforming the role from instructor to facilitator of human-machine collaboration. This means acquiring competencies in digital tools, adaptive pedagogy, sustainability, and continuous professional development.

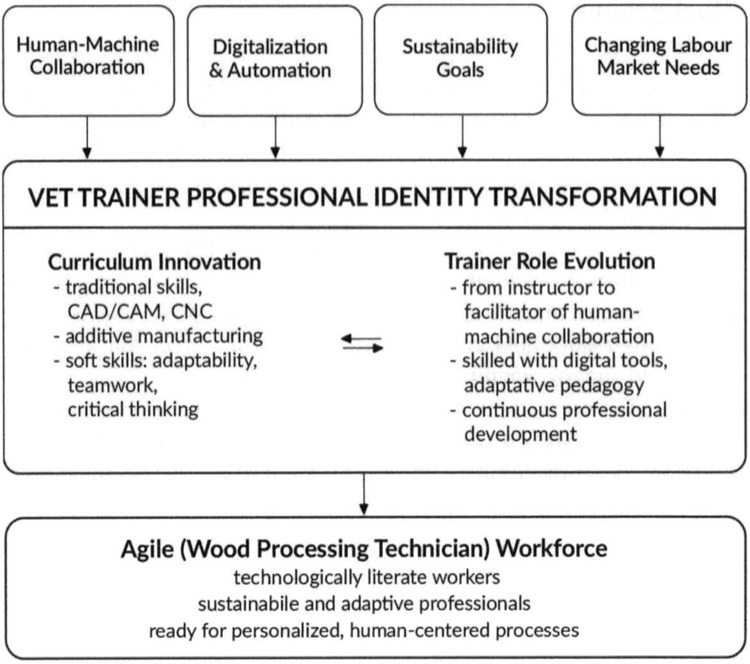

Fig. 5.1 The conceptual framework of this research. (Source: Authors' processing based on Maddikunta et al. (2022), McGrath et al. (2020), Demir et al. (2019), Schröder (2019); Hirschi (2018), Thoben et al. (2017))

The outcome of the framework (what the transformation aims to enable) is the third layer—the Agile Workforce in Wood Processing (Pejic-Bach et al., 2020; Bonekamp & Sure, 2015), consisting of technologically literate technicians, who should become sustainable and adaptive professionals capable of co-working with technologies like AI and robotics, and ready for personalized, human-centered production in I5.0.

Human-machine collaboration and AI. The integration of collaborative robotics and AI in woodworking processes demands a shift in the curriculum toward developing digital fluency and human-centered technology skills. Learners should figure out how to operate smart machines and how to work alongside them safely and efficiently. By introducing real-world scenarios involving collaborative robot applications and decision-support systems, the curriculum can empower future technicians to manage complexity and act as intelligent operators, rather than mere machine users (Fast-Berglund et al., 2016; Kadir et al., 2019).

Digitalization and automation. Digital workflows and automated systems are transforming how wood products are designed, fabricated, and tracked through production. A modern curriculum must include at least the basics of CNC programming, CAM software, digital design tools (CAD), and sensor-based quality monitoring. Learners should engage in project-based learning that simulates digitally controlled environments, with a focus on traceability, efficiency, and adaptability. This equips them with the ability to navigate and troubleshoot increasingly digitized production lines, making them agile contributors in smart manufacturing settings (Thames & Schaefer, 2016; Dragomir et al., 2016).

Sustainability goals. The growing emphasis on environmental responsibility influences both materials and processes in the wood industry. The curriculum should embed the core principles of circular economy, energy efficiency, and material optimization, while also promoting a mindset of care for natural resources. Topics such as wood waste recycling, low-impact finishes, and lifecycle assessment help learners make informed, sustainable decisions in their future roles (Bag et al., 2021; Niemann & Pisla, 2021).

Changing labor market needs. As the woodworking industry evolves, there is a rising demand for multi-skilled workers who can quickly adapt to technological change and new quality standards. The curriculum must therefore foster skills such as problem-solving, digital literacy, communication, and lifelong learning habits, enabling programs to respond to regional industry needs. Exposure to collaborative projects, customer-oriented thinking, and innovation tools enables learners to grow into proactive, resilient technicians able to meet the shifting demands of a dynamic labor market (Pfeiffer, 2018).

This research addresses a critical and timely need to align vocational education and training with the rapidly evolving demands of Industry 5.0. As the wood processing sector faces increasing pressure to integrate automation, digitalization, and sustainability, the ability of VET systems to prepare an agile and future-ready workforce becomes essential. However, this transition cannot be achieved through curriculum updates alone (Gessler & Howe, 2015; Billett, 2014). The professional identity of VET trainers—those who directly shape learners' skills, mindsets, and attitudes—must evolve to reflect their new roles as mediators between technology and human potential. Understanding and guiding this transformation is key to ensuring that vocational education remains relevant, effective, and empowering for both learners and industry.

While existing literature often emphasizes curriculum modernization or student outcomes, few studies explore how trainers must adapt professionally and pedagogically to lead this transition effectively. By examining how trainers can develop new competencies, attitudes, and teaching approaches to support human-machine collaboration and sustainability (among others), this study offers an insight into the systemic transformation required in VET. Its focus on the wood processing sector provides a grounded and under-researched case where tradition and innovation intersect in particularly challenging and instructive ways (Mulder, 2017).

5.3 Redefining the Professional Identity of VET Trainers

In this research, we propose rethinking the professional identity of VET trainers via a set of specific competencies that they need to develop in order to prepare an agile workforce in the context of Industry 5.0. In the VET system, the existence of a well-defined occupational standard (Mulder et al., 2007) and a predetermined curriculum offers limited flexibility for reshaping a training program. However, adjustments can be made in the training modules that are designed by the industrial actor(s) involved in the program, as well as in the upskilling of the trainers themselves (Andersson & Köpsén, 2015). One such module is foreseen for each year of study.

Based on the existing training modules and the competencies defined by the occupational standard, we proposed a set of "custom" training modules (those that can be defined by the industrial partner) with relevance for Industry 5.0. We then identified the competencies required by trainers to successfully deliver these modules. Our approach began with an analysis of the competencies outlined in the occupational standard—broken down into knowledge, skills, and attitudes—through the lens of their relevance to Industry 5.0. We considered the three core dimensions of Industry 5.0: technological, human-centric, and sustainability. Based on this analysis, we developed training modules designed to strengthen the competencies most aligned with these dimensions. This section details the steps taken.

According to the national curriculum approved by the Ministry of Education for grades 9 to 12 ("Standard de Pregătire Profesională—Thenician în Prelucrarea Lemnului," 2016), vocational learners are required to complete a series of training modules (see Table 5.1). These modules should lead to specific learning outcomes—knowledge, skills, and attitudes—foreseen by the national standard.

Each training module is designed to target a specific category of learning outcomes that eventually generate the competencies listed by the standard. However, we consider that a given module may contribute to the development of multiple learning outcomes, not just those it explicitly aims to address.

To identify the most relevant knowledge, skills, and attitudes from the list provided in the occupational standard, we evaluated each item in relation to the three key dimensions of Industry 5.0: technological, human-centric, and

Table 5.1 The training modules for the specialization "Wood Processing Technician"

	Training module
9th Grade	M9.1. Raw materials and technological materials in the wood industry M9.2. Drawing representation of simple wood products *M9.3. Local curriculum development*
10th Grade	M10.1. Timber drying M10.2. Timber cutting technology *M10.3. Local curriculum development*
11th Grade	M11.1. Raw materials and technological materials for the manufacturing of wood products M11.2. Drawing representation of finished wood products M11.3. Manufacturing processes of furniture and other finished wood products *M11.4. Local curriculum development* M11.5. Structures for finished wood products
12th Grade	M12.1. Technical documentation for finished wood products M12.2. Furniture decoration *M12.3. Local curriculum development* M12.4. Technological processes for obtaining superior wood-based semi-finished products

Source: Authors' processing

sustainability-related. A qualitative multicriteria assessment was applied, assigning scores from 1 to 10 for each dimension (10—direct, clear, consistent contribution with significant impact; 7–9—moderate to high contribution; 4–6—modest or indirect contribution; 1–3—weak or negligible contribution). By "contribution," we refer to the potential for the use of modern technologies (digitalization, automation, advanced manufacturing), the potential to support collaboration, teamwork, and human interaction in industrial processes, and the potential for efficient resource management and environmental impact reduction, including the use of sustainable materials and eco-friendly production processes.

Table 5.2 presents the most relevant knowledge, skills, and attitudes according to this evaluation, together with their scores. For brevity, we included only the top 10 items in each category, as the standard contains a comprehensive list of competencies. We retained the original numeric codes from the standard for reference and translated the competencies to English.

One of the key challenges faced by (local) industrial organizations involved in VET programs is the definition of adequate training modules. The prevailing tendency is to design content based on immediate economic needs. In contrast, our proposal is to define the content in alignment with the direction set by Industry 5.0, as outlined in the first layer of our framework—namely, the guiding vectors—and to incorporate the three core dimensions of I5.0: technological, human-centric, and sustainability.

We started from the existing structure of training modules across all 4 years of vocational education (Table 5.1) and analyzed them in terms of technological components that could be expanded, areas where sustainability is underrepresented, and opportunities to foster attitudes such as collaboration, hands-on learning, and initiative.

For each "Locally Developed Curriculum" module, we then proposed content that complements existing topics, anticipates future ones, and bridges the learners' experience with real-world industrial practices.

Each module was designed so that its content items address at least one element from each of the three I5.0 dimensions: technology, human-centeredness, and sustainability—see Table 5.3. We also drew inspiration from the "Top Ten Skills for the Future of Manufacturing" identified by the World Manufacturing Forum, as referenced in the European Commission's report Industry 5.0—Towards a sustainable, human-centric, and resilient European industry.

Trainers involved in VET programs come from economic and industrial sectors and are not professional educators by training. It is therefore essential that they possess a competence profile that enables them to effectively transfer knowledge and experience to students, to support the development of appropriate attitudes aligned with Industry 5.0, and to use modern training methods and tools.

To identify the competencies needed to effectively train a future agile workforce, we followed a multi-step process. We refer here to competencies that complement the trainers' existing technical background, rather than replacing it. First, we revisited the actual tasks of an industrial trainer involved in VET programs (the lead author of this study is directly engaged in such a program). A trainer should bridge

Table 5.2 The most relevant competencies for Industry 5.0, according to our multicriteria assessment

Knowledge Units	Techn.	Human	Sust.	Score
3.1.1. Artificial wood drying installations	10	8	9	9.0
1.1.1. Structure and macroscopic identification of wood	10	8	8	8.7
7.1.5. Designed product manufacturing technology description	10	6	8	8.0
8.1.6. Furniture assembly	10	9	5	8.0
9.1.3. Risk management in the furniture manufacturing process	10	10	4	8.0
1.1.4. Adhesives used in the wood industry	9	5	9	7.7
3.1.3. Cutting operations	9	10	4	7.7
5.1.1. Characteristic elements of technical drawing	9	10	4	7.7
6.1.3. Assembly of parts in the furniture structure	9	9	4	7.3
4.1.3. Wood properties	9	9	3	7.0
1.2.4. Selection of raw material assortments for manufacturing a given finished product	7	8	9	8.0
3.2.13. Measuring the dimensions of parts obtained by cutting and comparing them with those in the technological sheet	9	7	6	7.3
8.2.7. Using computers in mechanical processing with modern machinery (CNC machines)	10	7	7	8.0
4.2.15. Use of abrasive materials, depending on the nature of the substrate and the surface to be sanded	8	6	8	7.3
7.2.9. Indicating the occupational health and safety and fire safety regulations to be followed, according to the manufacturing technology of the designed product	7	9	8	8.0
4.2.13. Recognizing adhesives based on their characteristics and properties	9	7	8	8.0
2.2.4. Use of conventional symbols for representing raw materials and materials from the wood industry, for a given simple wood product	7	8	6	7.0
1.2.13. Use of specialized terminology to describe raw materials and technological materials used for the work task	8	7	6	7.0
6.2.4. Creating in a drawing the sketch of typical assemblies of wooden parts	8	7	6	7.0
5.2.13. Communication/Reporting of the results of professional activities performed	7	9	7	7.7
3.3.7. Taking responsibility for the quality of the cutting operations performed	9	9	9	9.0
1.3.7. Complying with occupational health and safety regulations, fire safety, and environmental protection when using raw materials and technological materials	9	9	9	9.0
4.3.5. Taking initiative in selecting raw materials and the finishing method for a given finished product	8	9	8	8.3
1.3.1. Awareness of the importance of recognizing wood species for the targeted qualification	8	7	8	7.7

(continued)

Table 5.2 (continued)

Knowledge Units	Techn.	Human	Sust.	Score
3.3.9. Taking responsibility for complying with occupational health and safety regulations, fire safety, and environmental protection when cutting parts from solid wood	9	9	9	9.0
9.3.12. Applying quality procedures responsibly in accordance with current regulations	8	8	8	8.0
7.3.5. Taking initiative in solving problems	8	9	8	8.3
8.3.8. Complying with the sequence of operations in the technological flow	8	7	8	7.7
6.3.9. Complying with environmental protection regulations	8	7	9	8.0
3.3.3. Complying with ergonomic requirements at the workplace	8	9	7	8.0

Source: Authors' own processing

Table 5.3 "Locally Developed Curriculum" training modules—our proposal

Learning module—Local curriculum development	Learning contents
9th Grade—M9.3. Introduction to Emerging Technologies for the Wood Industry Focus: Human Dimension	What is Industry 5.0 and how does it differ from Industry 4.0 (introduces the general concept, sets the context for the other units)—supports attitudes 1.3.1 and 7.3.5. Human-machine collaboration, social responsibility, creativity—supports attitudes 3.3.3, 3.3.7, 9.3.12. Recognition of wood species as an essential skill in the profession—supports knowledge 1.1.1, 4.1.3, skills 1.2.13. Responsibility, initiative, adherence to safety and environmental standards (Strengthens the foundation for OHS and fire safety standards)—supports attitudes 1.3.7, 3.3.9, 6.3.9. Case studies: examples of sustainable and ergonomic products—supports knowledge 4.1.3 and attitudes 3.3.3, 6.3.9. Evaluating posture ergonomics at the workbench—supports knowledge 1.1.1, 4.1.3, skills 1.2.4, attitudes 3.3.3, 1.3.1.
10th Grade—M10.3. Digitalization and Process Control in Wood Processing Focus: Technological Dimension	Introduction to the use of CNC equipment in the wood industry—supports knowledge 3.1.3, skills 8.2.7, attitudes 3.3.7, 3.3.9. Basics of computer-aided manufacturing (CAM)—supports knowledge 5.1.1, 6.2.4, skills 8.2.7. Integration of sensors in drying and cutting installations—supports knowledge 3.1.1, 3.1.3, attitudes 9.3.12. Using the computer to interpret technological sheets—supports knowledge: 7.1.5, skills: 5.2.13. CNC program simulation for a basic part—supports knowledge 3.1.3, skills 8.2.7, attitudes 3.3.7, 3.3.9.

(continued)

Table 5.3 (continued)

Learning module—Local curriculum development	Learning contents
11th Grade—M11.4. Responsible Manufacturing: Quality, Safety, and Process Optimization Focus: Sustainability Dimension	Risk management in the manufacturing process—supports knowledge 9.1.3, 9.3.12, skills 7.2.9. Health and safety standards and environmental protection in production—supports attitudes: 1.3.7, 3.3.9, 6.3.9, skills: 7.2.9. Alternative materials and wood waste recycling—supports knowledge: 1.1.1, 4.1.3, skills: 1.2.4, 1.2.13, attitudes: 6.3.9. Traceability and quality control at each stage—supports knowledge: 7.1.5, 9.1.3, skills: 5.2.13, attitudes: 9.3.12. Impact analysis of a furniture production technological flow—supports knowledge: 8.1.6, 7.1.5, skills: 3.2.13, 4.2.15, attitudes: 8.3.8.
12th Grade—M12.3. Collaborative Design and Innovation in Wood Products Focus: Technological, Human, and Sustainable Integration	User-centered design in wood product development—supports Knowledge: 5.1.1, Skills: 6.2.4, 2.2.4, Attitudes: 3.3.3, 7.3.5. Using sketches and digital prototypes (CAD) for cocreation—supports Knowledge: 5.1.1, 6.1.3, Skills: 8.2.7, 6.2.4, Attitudes: 4.3.5. Collaboration between departments: design, execution, marketing—supports Knowledge: 7.1.5, Skills: 5.2.13, Attitudes: 7.3.5, 9.3.12. Team project: from idea to final product—supports Knowledge: 7.1.5, 6.1.3, Skills: 1.2.4, 3.2.13, Attitudes: 8.3.8. Project presentation: communication, responsibility, feedback—supports Knowledge: 12.1.1, Skills: 5.2.13, Attitudes: 7.3.5, 9.3.12.

Source: Authors' processing

theoretical knowledge with industrial practice, support the development of both technical and transversal skills, and understand the curricular logic of vocational high schools.

We then used the standard framework of organizing competencies into knowledge, skills, and attitudes, and considered both the minimum pedagogical requirements for VET trainers and the technical skills relevant to each proposed training module (see Table 5.4). We also relied on key trainer competencies outlined in European Commission documents on Industry 5.0 (European Commission, 2022) and the Pact for Skills (European Commission, 2020) and followed three directions: (a) systemic thinking, human-centric approaches, and sustainability; (b) integration of digital and green technologies into training; and (c) the ability to collaborate with both schools and industry stakeholders. Finally, the resulting list was adapted to the local context, where trainers are not career educators but professionals from industry.

Table 5.4 The proposed trainer competencies

	Competencies
Knowledge	Current technologies in the wood industry: Modern machinery, CNC, digitalized processes; Types of innovative and sustainable materials; Basic concepts of CAD, CAM software, collaborative robots, and industrial process simulation. Integrated industrial processes with digitalization: Automation, IoT, traceability in production; Quality monitoring systems. Principles of industrial sustainability: Circular economy, waste reduction, energy efficiency. Health, safety, and environmental standards: Updated legal requirements; Industrial ergonomics. Foundations of pedagogy for VET training: Principles of experiential learning; Fundamentals of pedagogical communication; Techniques for explanation, demonstration, and feedback.
Skills	Communication with trainees: Clear explanation of tasks and technical concepts in a way that students can understand; Adapting content to the level and learning style of the students; Active listening and constructive feedback. Use of relevant equipment and software: CNC machines, cutting simulators, control interfaces; Basic CAD/CAM software. Applying practical teaching methods: Applied, collaborative projects, simulations. Managing unforeseen situations and risks; Adapting activities in real-time.
Attitudes	Openness to collaboration with other education stakeholders (teachers, VET coordinators, tutors). Acceptance of educational feedback. Empathy toward young people and understanding their needs. Responsibility for the quality of training: Respecting educational objectives, not just industrial ones. Attention to inclusion and equity: Adapting language and examples to the level of all learners.

Source: Authors' processing

5.4 Discussion, Contributions, and Impact

The study highlights how VET trainers' responsibilities should shift from mere knowledge transmission to the facilitation of collaborative learning between humans and machines. Based on our framework, three dimensions of the new professional identity of trainers (Köpsén, 2014) are considered.

The first is the shift from traditional instructors to human-machine collaboration facilitators. Trainers should transition from focusing on manual craftsmanship to mentoring learners in a technology-enhanced learning environment. This includes hands-on training with advanced digital tools and fostering an understanding of the synergy between human intuition and automation.

The second dimension covers the essential competencies for VET Trainers in Industry 5.0: advanced digital skills (familiarity with CAD/CAM software, CNC programming, industrial process simulation), pedagogical adaptability (the ability to integrate traditional craftsmanship with digital workflows and problem-solving methodologies), and sustainability expertise (incorporating eco-friendly materials and energy-efficient techniques into training modules).

The third dimension is embedding sustainability in the VET programs. As sustainability becomes a central pillar of Industry 5.0, VET trainers should educate learners on material innovations (e.g., use of engineered wood, recycled materials, and eco-friendly adhesives), energy-efficient manufacturing (optimization of production processes to reduce waste and energy consumption), and circular economy principles (designing products with longevity, reusability, and recyclability in mind).

This framework was developed specifically for the wood processing industry within the Romanian VET system, where industrial partners design locally developed curriculum modules. The Romanian context provides a relevant case study due to its traditional craft-based approach to wood processing education, which now requires integration with Industry 5.0 principles.

While requiring empirical validation, this study contributes to both academic literature and practical applications in the following ways. First, the framework for VET adaptation to Industry 5.0 can serve as a roadmap for educators, policymakers, and industry stakeholders in aligning vocational training with technological and sustainability trends. Second, we provide a structured approach for updating VET programs to meet future labor market demands. Third, the study enables targeted professional development strategies to equip VET trainers with the necessary digital, pedagogical, and sustainability skills.

5.5 Conclusions

As Industry 5.0 reshapes traditional industries, vocational education must evolve to foster an agile, technology-driven, and sustainability-conscious workforce. This study, grounded in the Romanian wood processing VET context, argues that by reshaping the professional identity of VET trainers, vocational education can bridge the gap between traditional craftsmanship and digital innovation.

Transforming the professional identity of VET trainers enhances businesses' sustainability and resilience by equipping trainers with the skills and mindset necessary to adapt to evolving industry demands. By focusing on interactive learning, modern tools such as 2D/3D drawing and CNC simulation, and integrating sustainability principles into the curriculum, trainers can better prepare students for the challenges of a rapidly changing market. This approach fosters a culture of continuous improvement, innovation, and collaboration, ensuring that businesses involved in VET are more agile, capable of responding to emerging trends, and committed to sustainable practices, ultimately strengthening their long-term viability and competitiveness.

Future research will explore the implementation of this framework in the Romanian wood processing sector and other VET contexts, measuring its impact on learner employability and industry adaptability.

References

Alam, A., & Mohanty, A. (2023). Educational technology: Exploring the convergence of technology and pedagogy through mobility, interactivity, AI, and learning tools. *Cogent Engineering, 10*(2), 2283282. https://doi.org/10.1080/23311916.2023.2283282

Andersson, P., & Köpsén, S. (2015). Continuing professional development of vocational teachers: Participation in a Swedish National Initiative. *Empirical Research in Vocational Education and Training, 7*, 7. https://doi.org/10.1186/s40461-015-0019-3

Autor, D. H. (2015). Why are there still so many jobs? The history and future of workplace automation. *Journal of Economic Perspectives, 29*(3), 3–30. https://doi.org/10.1257/jep.29.3.3

Bag, S., Gupta, S., & Kumar, S. (2021). Industry 4.0 adoption and 10R advance manufacturing capabilities for sustainable development. *International Journal of Production Economics, 231*, 107844. https://doi.org/10.1016/j.ijpe.2020.107844

Billett, S. (2014). The standing of vocational education: Sources of its societal esteem and implications for its enactment. *Journal of Vocational Education & Training, 66*, 1–21. https://doi.org/10.1080/13636820.2013.867525

Bonekamp, L., & Sure, M. (2015). Consequences of industry 4.0 on human labour and work organisation. *Journal of Business and Psychology, 6*, 33–40.

Cedefop. (2020). *Vocational education and training in Europe, 1995–2035. Scenarios for European vocational education and training in the 21st century* (Cedefop Reference Series No 114). Publications Office. http://data.europa.eu/doi/10.2801/794471

Chankseliani, M., & Anuar, A. M. (2019). Cross-country comparison of engagement in apprenticeships: A conceptual analysis of incentives for individuals and firms. *International Journal for Research in Vocational Education and Training, 6*, 261–283. https://doi.org/10.13152/IJRVET.6.3.4

Demir, K. A., Döven, G., & Sezen, B. (2019, June 21–23). Industry 5.0 and human-robot coworking. In *Procedia computer science, 3rd world conference on technology, innovation, and entrepreneurship industry 4.0 focused innovation, technology, entrepreneurship and manufacture* (Vol. 158, pp. 688–695). https://doi.org/10.1016/j.procs.2019.09.104

Dragomir, M., Banyai, D., Dragomir, D., Popescu, F., & Criste, A. (2016). Efficiency and resilience in product design by using morphological charts. *Energy Procedia, 85*, 206–210. https://doi.org/10.1016/j.egypro.2015.12.218

European Commission. (2020). *The pact for skills: Mobilizing all partners to invest in skills.* https://pact-for-skills.ec.europa.eu/index_en

European Commission. (2022). *Industry 5.0, a transformative vision for Europe* (ESIR, Policy Brief Report No. 3). Publications Office of the European Union. https://www.horizon-europe.gouv.fr/sites/default/files/2022-01/industry-5-0-pdf-5324.pdf

Fast-Berglund, Å., Palmkvist, F., Nyqvist, P., Ekered, S., & Åkerman, M. (2016). Evaluating Cobots for final assembly. In *Procedia CIRP, 6th CIRP Conference on Assembly Technologies and Systems (CATS)* (Vol. 44, pp. 175–180). https://doi.org/10.1016/j.procir.2016.02.114

Gessler, M., & Howe, F. (2015). From the reality of work to grounded work-based learning in German vocational education and training: Background, concept and tools. *International Journal for Research in Vocational Education and Training, 2*, 214–238. https://doi.org/10.13152/IJRVET.2.3.6

Hirschi, A. (2018). The fourth industrial revolution: Issues and implications for career research and practice. *The Career Development Quarterly, 66*, 192–204. https://doi.org/10.1002/cdq.12142

Kadir, B. A., Broberg, O., & da Conceição, C. S. (2019). Current research and future perspectives on human factors and ergonomics in industry 4.0. *Computers and Industrial Engineering, 137*, 106004. https://doi.org/10.1016/j.cie.2019.106004

Köpsén, S. (2014). How vocational teachers describe their vocational teacher identity. *Journal of Vocational Education & Training., 66*, 194–211. https://doi.org/10.1080/13636820.2014.894554

Maddikunta, P. K. R., Pham, Q.-V., Prabadevi, B., Deepa, N., Dev, K., Gadekallu, T. R., Ruby, R., & Liyanage, M. (2022). Industry 5.0: A survey on enabling technologies and potential applications. *Journal of Industrial Information Integration, 26*, 100257. https://doi.org/10.1016/j.jii.2021.100257

Marope, P. T. M., Chakroun, B., & Holmes, K. P. (2015). *Unleashing the potential: Transforming technical and vocational education and training*. UNESCO Publishing. https://doi.org/10.54675/HXES1603

McGrath, S., Ramsarup, P., Zeelen, J., Wedekind, V., Allais, S., Lotz-Sisitka, H., Monk, D., Openjuru, G., & Russon, J.-A. (2020). Vocational education and training for African development: A literature review. *Journal of Vocational Education and Training, 72*, 465–487. https://doi.org/10.1080/13636820.2019.1679969

Mulder, M. (Ed.). (2017). *Competence-based vocational and professional education: Bridging the worlds of work and education, technical and vocational education and training: Issues, concerns and prospects*. Springer International Publishing. https://doi.org/10.1007/978-3-319-41713-4

Mulder, M., Weigel, T., & Collins, K. (2007). The concept of competence in the development of vocational education and training in selected EU member states: A critical analysis. *Journal of Vocational Education and Training, 59*, 67–88. https://doi.org/10.1080/13636820601145630

Nahavandi, S. (2019). Industry 5.0 – A human-centric solution. *Sustainability, 11*, 4371. https://doi.org/10.3390/su11164371

Niemann, J., & Pisla, A. (2021). Life cycle information support. In J. Niemann & A. Pisla (Eds.), *Life-cycle management of machines and mechanisms* (pp. 57–73). Springer International Publishing. https://doi.org/10.1007/978-3-030-56449-0_6

Pejic-Bach, M., Bertoncel, T., Meško, M., & Krstić, Ž. (2020). Text mining of industry 4.0 job advertisements. *Journal of Vocational Education and Training, 50*, 416–431. https://doi.org/10.1016/j.ijinfomgt.2019.07.014

Pfeiffer, S. (2018). The 'future of employment' on the shop floor: Why production jobs are less susceptible to computerization than assumed. *International Journal for Research in Vocational Education and Training, 5*, 208–225. https://doi.org/10.13152/IJRVET.5.3.4

Pilz, M., & Wiemann, K. (2021). Does dual training make the world go round? Training models in German companies in China, India and Mexico. *Vocations and Learning, 14*, 95–114. https://doi.org/10.1007/s12186-020-09255-z

Schröder, T. (2019). A regional approach for the development of TVET systems in the light of the 4th industrial revolution: The regional association of vocational and technical education in Asia. *International Journal of Training Research, 17*, 83–95. https://doi.org/10.1080/14480220.2019.1629728

Standard de Pregătire Profesională – Tehnician în Prelucrarea Lemnului. (2016). https://www.edu.ro/sites/default/files/_fișiere/Invatamant-Preuniversitar/2016/liceal/standarde/SPP_niv4_Th_proiectant_produse_finite_din_lemn.pdf

Susskind, R. E. (2015). *The future of the professions: How technology will transform the work of human experts*. Oxford University Press.

Thames, L., & Schaefer, D. (2016). Software-defined cloud manufacturing for industry 4.0. In *Procedia CIRP, The 6th International Conference on Changeable, Agile, Reconfigurable and Virtual Production (CARV2016)* (Vol. 52, pp 12–17). https://doi.org/10.1016/j.procir.2016.07.041

Thoben, K.-D., Wiesner, S., & Wuest, T. (2017). 'Industrie 4.0' and smart manufacturing – A review of research issues and application examples. *International Journal of Automation Technology, 11*, 4–16. https://doi.org/10.20965/ijat.2017.p0004

Xu, X., Lu, Y., Vogel-Heuser, B., & Wang, L. (2021). Industry 4.0 and industry 5.0 – Inception, conception and perception. *Journal of Manufacturing Systems, 61*, 530–535. https://doi.org/10.1016/j.jmsy.2021.10.006

Chapter 6
Quality, Energy Consumption, and Economic Growth: A Quantitative Approach

Viorela-Ligia Văidean and Alexia-Raluca Turceniuc

Abstract Is economic growth fundamentally at odds with our planet's uncertain future? A groundbreaking global analysis offers a roadmap to reconcile the two. The study examines the dynamic interplay between economic growth, energy consumption, and environmental sustainability across 185 countries from 2000 to 2022. The primary aim is to analyze how these factors interact amidst the intensifying global climate crisis and sustainability commitments, focusing on the trade-offs between energy use and economic expansion. We employ a panel fixed effects model to quantify the effects of the energy consumption, greenhouse gas emissions, and human development on the GDP per capita. Our main contributions indicate: (1) establishing human development, as captured by the Human Development Index (HDI), as a strong determinant of economic growth; (2) demonstrating that while per capita energy consumption positively boosts the GDP per capita, total energy consumption and electricity generation have negative effects, suggesting inefficiencies in energy utilization; and (3) identifying a nonlinear relationship between the GHG emissions and growth, where emissions initially hinder the economic performance but contribute positively when considered alongside energy variables. These findings underscore the critical need for strategic energy management and policy intervention, particularly in the emerging economics, advocating for enhanced energy efficiency, renewable energy integration, and sustainable growth strategies aligned with environmental goals.

Keywords Economic growth · Energy consumption · Environmental quality · Greenhouse gas emissions · Sustainable development

V.-L. Văidean (✉) · A.-R. Turceniuc
Faculty of Economics and Business Administration, Babeş-Bolyai University, Cluj-Napoca, Romania
e-mail: viorela.vaidean@econ.ubbcluj.ro

A. L. Negrusa, M. M. Coroş (eds.), *Innovation, Sustainability, and Growth in a VUCA Environment*, Springer Proceedings in Business and Economics,
https://doi.org/10.1007/978-3-032-11639-0_6

6.1 Introduction

This chapter empirically explores the impact of energy consumption and environmental quality on economic growth in a panel of 185 countries covering the period 2000–2022. Understanding these relationships is crucial as the balance between environmental sustainability and economic prosperity becomes increasingly significant for governments and for the well-being of nations. Greenhouse gas emissions are known to drive climate change, which in turn may affect economic conditions and human well-being through extreme weather events, loss of biodiversity, and health impacts. NASA predicts that the polluting emissions could raise the Earth's temperature between 2 °C and 6 °C by the end of the century (Riebeek 2010). Conversely, energy consumption is often directly linked to economic growth, providing the indispensable power for industries, infrastructure, and technological improvements, but it also poses significant environmental challenges due to its association with the dependence on fossil fuels and pollution. Triantafyllidou and Polychronidou (2025) empirically demonstrate the differentiated effects of fossil fuels and renewable energy consumption, positively contributing to economic growth while mitigating the emissions across EU nations. Thus, the analysis of these factors together will provide insights into strategies that aim to mitigate environmental damage caused by climate change, while promoting economic and human progress.

The novelty of this research lies in its scope and integrative approach. The choice of leveraging an extended dataset that spans nearly all the nations worldwide provides a panoramic data-driven investigation into the complex ties between energy use, environmental sustainability, and economic growth. What sets this work apart lies in drawing from two decades of data across almost every nation on Earth, presenting an imperative empirical work in this field. Unlike previous research that often isolates single variables and regions, our approach focuses on the overall interaction of all the factors, rather than the individual impacts of energy use and environmental quality on the economic growth. Nevertheless, our comparative approach captures both developed and developing nations, highlighting the challenges and opportunities that affect sustainable growth in this interconnected world. The purpose is to understand how these nations grow economically and also sustainably and to analyze the relationships between them. By applying a rigorous panel fixed effects model, which helps us control for unseen differences across countries, we identify patterns that hold up globally. Our main objectives are guided by a strong dedication to advancing sustainable growth, setting out four objectives. First, we seek to unpack how energy consumption and greenhouse gas emissions influence a nation's economic performance. Second, the study examines the role of human development, using the Human Development Index as a key factor, indirectly uncovering how improvements in education, living standards, and health might contribute to the expansion of economies. Third, it seeks to identify development strategies that support economic and human progress while improving the environmental quality and reducing pollution. Lastly, the research is designed to provide solid insights that might be helpful for policymakers, especially in the less developed nations that are building their economic foundations. The central hypothesis of this work states that while energy use continues to drive economic development, the emissions it produces create severe challenges for a sustainable environment.

The importance of this study is underscored by the global commitment to a number of 17 sustainable development goals introduced by the United Nations (United Nations, 2017) and the Paris Agreement (UNFCCC, 2015), both aiming to address critical climatic challenges and encompass a set of goals that are sustainable and reliable for anyone. The world's top emitters, including China, India, the USA, Japan, and Russia, collectively contribute a significant portion of CO_2 emissions, with China leading at 29%, followed by the USA at 16%. These nations hold the power to reduce global emissions but ultimately, their efforts are hindered by the fact that their emissions are closely tied to their level of energy production, which drives their economies. Consequently, there is reluctance to implement measures that could potentially harm economic growth, highlighting the sensitive balance between environmental matters and economic stability. By examining the interplay between emissions, energy usage, and the prosperity of nations, we can identify patterns and propose strategies that support both economic growth and a sustainable environment. This is particularly relevant for developing and less developed nations which are striving to enhance their economic status without compromising their environmental future.

To achieve our research objectives, we employ a multivariate data approach using panel fixed effects estimators. Simple regressions help us understand the direct connection between independent variables, such as the impact of human development on economic growth. The extended version of the fixed effects model analysis will allow us to consider the combined effect of multiple variables, offering a more comprehensive view of how energy consumption and emissions altogether can influence economic growth. This methodology enables us to quantify the connections between the proxies and assess the relative importance of each factor, providing a better understanding of the dynamics at play. We use data from reputable sources such as the World Bank, the UNDP (Human Development Reports), the International Energy Agency (IEA), and the Global Carbon Budget. This comprehensive dataset will include indicators such as levels of GDP per capita, HDI scores, greenhouse gas emissions, CO_2 emissions, energy usage and generation metrics, covering a wide range of developed and developing countries and allowing for a comparative analysis across different contexts.

The research is structured as follows: the first part provides an overview of the theoretical framework regarding the impact of greenhouse gas emissions and energy consumption on the economic growth of nations, including a review of the literature and some key concepts. The second part details the process of data collection, including the sources of our data and the techniques used for our regression models. Moving forward, the third part presents the results, interpreting the statistical findings in the context of economic theories and development strategies. This part will also explore the policy implications of our findings, suggesting measures that can help balance economic growth with environmental sustainability in both developed and developing nations. Furthermore, the challenges and limitations of our study associated with the analyzed data will also be discussed. Finally, the chapter embraces a summary of the main conclusions and discusses some potential avenues for future research, emphasizing the importance of continuous monitoring and adaptive policies in the face of changing environmental and economic conditions.

6.2 Literature Review

Exploring the Relationship Between Climate Change and Other Influencing Factors of Economic Growth

Climate change is one of the most difficult ongoing concerns that humanity is ever going to tackle, and it refers to significant and enduring shifts in worldwide climate trends over extended periods, often spanning decades to millennia. At its core, climate change signifies a departure from historical norms of the climate, driven primarily by human-induced activities. Climate denotes the prolonged period, typically spanning at least 30 years, during which the average temperature, patterns of humidity, and rainfall are observed regionally and globally across all the seasons, yearly, for decades. In the research paper of Patlins et al. (2020), it is highlighted that the visualization of our main issue, climate change, "characterized by the first epistemological uncertainty," being a relatively new concern, a socio-ecological event rising in the science and education. Moreover, the study of climate change is pivotal for fostering environmental preservation and economic growth.

Across many countries worldwide, governments are questioning whether there is a possibility to maintain an equilibrium between the level of carbon dioxide emissions and the rapidly rising development of the economies. Energy demand continues to grow yearly for more than half a century, as the population is increasing and the living standard is steadily improving. Our existing energy infrastructure is a primary contributor to global warming, responsible for approximately 75% of total greenhouse gas emissions (IEA, 2023). Furthermore, while clean energy solutions are advancing, the ongoing use of fossil fuels and deforestation present significant obstacles to achieving the reduction of emissions. This underscores the need for urgent action and a comprehensive approach to address all aspects of the energy transition. Transforming how we generate and consume energy is crucial, as the world's ability to attain the climate goals set by the Paris Agreement of the United Nations Framework Convention on Climate Change, aiming to reach net zero emissions by the midpoint of the century, starting from 2020.

According to the Intergovernmental Panel on Climate Change (IPCC, 2022), it is crucial to prevent global temperatures from increasing by over 1.5 °C compared to pre-industrial thresholds in order to prevent catastrophic outcomes. The report emphasizes the substantial difference between a 1.5 °C and 2 °C increase in warming. The IPCC suggests that focusing efforts on restricting the rise in global temperature to 1.5 °C could yield several meaningful benefits by 2050. These include a substantial reduction in the number of individuals vulnerable to the dual challenges of poverty and climate change. By implementing effective measures, we could potentially reduce the vulnerability of millions of people around the world. Additionally, such efforts could safeguard approximately 10 million individuals from the adverse impacts of rising sea levels, offering crucial protection to coastal communities. Having set a strong emphasis on the urgency of taking decisive action to avoid a climate catastrophe, there appears to be a clear imperative for governments to drastically reduce greenhouse gas emissions by 2030, aiming to halve them compared to 2010 levels. This ambitious target is essential for keeping global temperatures within the 1.5 °C threshold and preventing irreversible damage to the planet's ecosystems. In the year 1950, the world emitted a quantity of 6 billion tons of CO_2 emissions (Ritchie & Roser, 2020). This quantity of polluting emissions had almost quadrupled by 1990, and it had exceeded more than 20 billion tons. Polluting emissions have continued to escalate swiftly since then, now exceeding 35 billion tons each year. While growth has experienced a downward trend in recent years, emissions have yet to reach their highest point. Consequences would negatively affect human health, a long-living safe environment, and the development of the population, and the economy will be jeopardized, and its expected pace of growth will be affected. In light of these findings, it is crucial that governments worldwide prioritize immediate action to shift the path toward a more sustainable and climate-resilient future. Failure to act swiftly will only exacerbate the challenges we face, forcing us to rely on costly technologies that may have adverse implications for human rights and will increase the existing inequalities.

Although some revolutionary energy-developed technologies, such as photovoltaic panels, electric cars, heating pumps, and batteries, have seen rapid growth, there is still an opportunity to limit the rise in global temperatures. A significant step toward raising awareness of climate change is the Paris Agreement (2015), which is an accord signed by 196 countries aimed at reducing and combating these changes and keeping global warming under 2 °C. To achieve this goal, the signatory countries have committed to significantly reduce greenhouse gas emissions and support their adjustment to climate change. However, significant and faster progress on a larger scale is essential, as highlighted by the IEA analysis (2023). Achieving this goal requires increased international collaboration and a stronger commitment from policymakers to shape a safer environment and a sustainable future.

The Evolution of Greenhouse Gas Emissions

Throughout the diurnal cycle, the radiant energy from the sun permeates Earth's atmosphere, initiating a process where the surface of the planet absorbs and stores heat. Conversely, during the nocturnal phase, this stored heat is gradually released back into the atmosphere as the surface cools. However, this thermal equilibrium is disturbed by greenhouse gases, which intercept and retain a portion of this outgoing heat, similar to the function of the glass panes in a greenhouse. This natural phenomenon, known as the greenhouse effect, acts as a regulatory mechanism and maintains the temperature on Earth at an average of approximately 14 °C, a pivotal condition essential for sustaining life on our planet. In the absence of the greenhouse effect, Earth's temperatures would experience a major downfall, potentially reaching as low as −18 °C, rendering the environment unsuitable for the flourishing of life in its forms. Nevertheless, human activities have significantly disrupted this sensitive balance by unleashing a burst of greenhouse gas emissions into the atmosphere.

Since the event of the Industrial Revolution, human endeavors have released a large volume of polluting emissions into the atmosphere, primarily through the combustion of fossil fuels and also through industrial activities. Over the past century, this emission rate has shown an upward trend in global temperatures. Notably, the past three decades have witnessed an alarming acceleration in this trend, with the current global temperatures registering the highest levels recorded in our history. Developed nations are responsible for approximately 53% to 61% of the increases in global air temperature and climate change, while developing nations contribute around 39% to 47%. The primary cause of the spatial variability in these changes from 1850 to 2005 is linked to greenhouse gas emissions from developed countries. Although there are still uncertainties in climate models and external forcings, polluting emissions from developed nations have become the main contributors to the climate changes observed in the twentieth century (Wei et al., 2015).

Merely focusing on individual areas such as electricity, deforestation, transportation, and the production of food is inadequate. Even within the energy sector, which is a large contributor of nearly three-quarters of the emissions, there is no easy remedy to solve the ongoing matter. The report of United Nations Environment Programme (2017) also indicated that emissions growth is particularly responsive in developed countries that are heavily dependent on fossil fuels to drive their economic growth and meet energy demands. By 2040, emissions from OECD countries are projected to reach approximately 13.8 billion metric tons, while non-OECD nations are expected to release around 29.4 billion metric tons. This represents an increase of about 8% and 51%, respectively, compared to the 2012 emissions levels. If we were to achieve complete decarbonization of our electricity generation, additional efforts would be necessary to help the transition of heating and transportation to electric sources. Moreover, challenges persist in sectors like shipping and aviation, where viable low-carbon technologies are not yet available (Fig. 6.1).

According to the International Energy Agency (CO_2 Emissions in 2022), the assessment of CO_2 emissions encompasses emissions from all fossil fuel applications related to energy objectives, encompassing the burning of non-renewable waste, alongside emissions originating from industrial operations like cement production, steel and iron manufacturing, and also chemical processing. Calculations of emissions from industrial processes rely on the most recent production figures for steel and iron, cement, aluminum, and other chemicals. Additionally, emissions

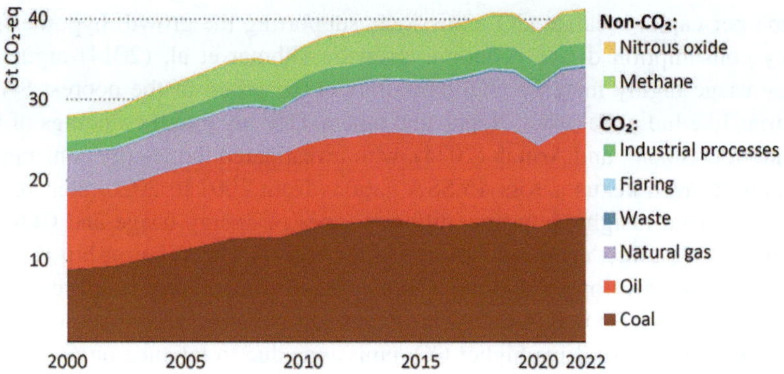

Fig. 6.1 Global energy-related greenhouse gas emissions, 2000–2022. (Source: Flaring emissions are from IEA analysis based on the World Bank Global Gas Flaring Reduction Programme)

resulting from the combustion of flared gases are factored into estimations of energy spread globally, associated with greenhouse gas emissions.

Non-CO_2 greenhouse gas emissions account for fugitive emissions emanating from the supply of oil, coal, and gas. Furthermore, assessments include emitted methane and nitrous oxide linked to energy combustion, with emission factors tailored to specific regions and end uses. The conversion of non-CO_2 greenhouse gas emissions into a distinctive manner of equivalent amounts utilizes a global warming potential over a span of 100 years, employing the global warming potential numbers which are assigned to methane and nitrous oxide.

Energy-Growth Relationship

The relationship that arises from the usage of energy and economic growth, namely the energy-growth nexus, has been extensively studied through four primary hypotheses. These theories are represented by the growth hypothesis, the conservation hypothesis, the feedback hypothesis, and the neutrality hypothesis. They offer a distinct perspective on the causality and interplay between energy usage and economic growth, with varying implications for policy in both developed and developing countries.

The growth hypothesis emphasizes that increased energy consumption drives economic growth. This view perceives energy as a critical input in the production processes, with higher energy use leading to enhanced productivity and economic expansion. Empirical studies supporting this hypothesis are more prevalent in developing countries, where energy infrastructure is expanding, and the marginal return on energy investment is significant. The findings of Gershon et al. (2024) underscore that in African nations, increased energy consumption is a relevant driver of economic expansion, and a contributor to heightened emissions. Salahuddin et al. (2015) found that in Gulf Cooperation countries, both energy demand and income per capita escalate CO_2 emissions, supporting the growth hypothesis that energy consumption drives economic growth. Akhmat et al. (2014) argued that energy usage highly increase CO_2 emissions in the group of the poorest SAARC countries like India, Pakistan, Nepal, and others. Delving into the findings of Kulu, Karakara, Dasmani, and Armah (2024), who investigated the energy consumption-economic growth nexus across 46 SSA nations from 2007 to 2020, their research provides crucial insights into how different types of energy usage and CO_2 emissions positively affect economic growth in the region. This relationship is particularly evident in developing regions where energy infrastructure is still expanding. Agusalim et al. (2025) emphasize that targeting economic growth in low-income countries continues to drive higher CO_2 emissions due to reliance on conventional energy sources. The conservation hypothesis argues that the economic development of nations results in increased energy consumption. Among their findings, Lee and Chang (2007) present evidence indicating that in a sample of 18 developing countries, economic growth drives energy consumption. This suggests that as these nations' GDP increases, their energy usage correspondingly rises, highlighting the dependency of energy needs on economic progress in developing regions. Furthermore, their study reveals the prevalence of the feedback hypothesis in a sample of 22 developed countries. This perspective of conservation is grounded in the notion that as economies grow, their energy needs expand due to greater industrial activity, higher standards of living, and increased demand for goods and services. Studies validating this hypothesis are frequently observed in developed countries, where mature economies exhibit a higher demand for energy as a result of economic growth.

The feedback hypothesis highlights a bidirectional bond between energy usage and economic growth, implying that they mutually reinforce each other. This hypothesis finds substantial empirical support across various contexts, highlighting a dynamic interaction where energy consumption spurs economic growth, and in turn, this economic development increases energy consumption. In contrast to the growth hypothesis, Tran et al. (2022) analyze 26 OECD countries and confirm the feedback hypothesis, showing that energy consumption and economic growth mutually influence each other. Lu, Ma, and Hu (2024) reassess the link between energy consumption and economic growth by focusing on economic growth rate forecasting, showing that variations in energy consumption can substantially affect economic trends. The study of Magazzino et al. (2021) offers insights into the temporal dynamics of the energy-growth nexus in Italy by employing a wavelet analysis. The results imply a bidirectional causality evident in the short term, while in the long run, the causal flow predominantly runs from economic growth to energy consumption. The situation for BRICS nations analyzed by Haseeb et al. (2018) states that an upward trend in GDP per capita in South Africa, Russia, and Brazil, leads to a reduction in CO_2 emissions, while in India and China, it results in higher CO_2 emissions. Similarly, energy consumption in Brazil, India, and China significantly raises CO_2 emissions. On the other hand, energy consumption in Russia has contributed to reducing CO_2 emissions, and in South Africa, it has no significant impact. Zhao et al. (2023) reveal a bidirectional relationship between energy use and economies at the national level in China, highlighting the importance of region-specific policies to support environmental sustainability. Moreover, a recent study of Lefatsa and Nubong (2025) confirms the feedback hypothesis in the SADC region, reaching a bidirectional Granger causal link between economic growth and energy use for a timeframe of almost 40 years, showing that growth drives the energy consumption and vice versa.

The neutrality hypothesis posits that there is no statistically significant connection between energy usage and economic growth. The study of Akhmat et al. (2014) also determined that methane emissions in Pakistan, Bangladesh, and Sri Lanka, along with extreme temperatures in India and Sri Lanka, have no influence on energy consumption, aligning with the neutrality hypothesis. This view implies that changes in the consumption of energy do not affect economic growth and vice versa. Eyuboglu and Uzar (2022) provide robust evidence affirming the neutrality hypothesis between renewable energy consumption and economic growth across 15 emerging economies for the timeframe 1990–2015, with no significant causality detected, but with negative economic shocks reducing the quantity of energy consumed in some nations. The study of Salari et al. (2021) examines the ties between energy and economic growth in the United States, finding evidence that supports neutrality, implying that reductions in energy use do not necessarily hinder economic expansion. Empirical support for this hypothesis is less common but has been identified in certain contexts, often in highly diversified economies where energy is not a limiting factor for growth.

While the feedback hypothesis garners the most empirical validation, suggesting a bidirectional and reinforcing association between energy usage and economic growth, the relevance of each hypothesis varies by country and developmental stage. Understanding these nuances is critical for crafting effective energy and economic policies tailored to specific national contexts.

6.3 Data and Methods

Description of Variables

The sole dependent variable is represented by the Gross Domestic Product per capita (Y) which signifies the average economic output generated per individual in a certain region. This proxy, expressed in international USD, is sourced from the World Databank and serves as an indicator for the economic performance across different geographical regions.

The independent variables in this analysis include the Human Development Index (K/L), Primary Energy Consumption (EC), Primary Energy Consumption per Capita (ECC), and the Electricity Generation (EG). The HDI offers a comprehensive measure of a nation's development status by encompassing dimensions such as education, level of income, and life expectancy, with data collected from the United Nations Development Programme (UNDP). Primary Energy Consumption (EC) captures the overall energy demand, while Primary Energy Consumption per Capita (ECC) provides insights into energy use at the individual level, highlighting patterns of energy efficiency and consumption behaviors. Electricity Generation (EG) reflects the total energy production within a region, indicative of the scale of its energy infrastructure. Greenhouse Gas emissions (GHG) and Carbon Dioxide emissions per capita (CARBON) are employed as control variables within this study. GHG assesses the environmental impact of economic activities, offering a broad perspective on ecological sustainability. CO_2 emissions per capita further refine this focus by examining the carbon footprint of a person, emphasizing individual contributions to environmental degradation. These control variables are crucial for isolating the direct impact of energy consumption on economic growth from the broader environmental consequences. Mitigating potential interfering effects and enhancing the robustness of the analysis. This approach ensures that the findings accurately reflect the interdependency between energy and economic development, allowing for complex outcomes of economic expansion on one hand, and sustainable development constraints on the other hand.

The sample includes 185 countries from all around the world, and the analyzed period is 2000–2022, the most recent data available at this moment. Further, we intend to find the nexus between the environmental quality, energy usage, and economic growth (Table 6.1).

Table 6.1 Description of variables

Acronym	Variable	Definition	Measurement	Source
Y/L	Gross Domestic Product (GDP) per capita	Income per capita	Purchasing power parity, international US dollars	World Bank Development Indicators
K/L	Human development index HDI	Assesses the development of a country with a long and healthy lifestyle and a satisfactory quality of living	Ranges between 0 and 1	United Nations Development Program (UNDP) Human Development Reports
EC	Primary energy consumption	Overall primary energy usage	Terawatt-hours	Energy Institute—Statistical Review of World Energy
ECC	Per capita primary energy consumption	Overall primary energy usage per person	Kilowatt-hours per person	Energy Institute—Statistical Review of World Energy
EG	Electricity generation	Total electricity generated from all sources	Terawatt-hours	Energy Institute—Statistical Review of World Energy
GHG	Environmental quality	Greenhouse gas emissions	Millions of metric tons of CO_2 equivalents	International Energy Agency (IEA)
CARBON	Environmental quality	Per capita carbon dioxide (CO_2) emissions from fossil fuels and industry. Land-use change is not included	Metric tons per capita	Global Carbon Budget

Note: Authors' own elaboration

Descriptive Statistics

According to the summary statistics presented below, the pinnacle level of the GHG is captured in China, at a level of 11,314 million tons of CO_2 equivalents, being the largest emitter of greenhouse gas emissions, mainly due to its substantial industrial activity and reliance on coal for energy. The lowest level of 0.55 is scored in a developing country, Eritrea, and it shows positive results in terms of sustainability, but on the other hand, it highlights some challenges when it comes to balancing economic development with environmental responsibility. Over the years included in the study, the average level of GHG is approximately 329.8029. Moreover, we may notice that the GHG values usually have the tendency to increase during the years from the specified period, but with some exceptions, such as Libya, Kazakhstan, and Venezuela, where high fluctuations are visible, followed by a descending trend, which suggests a potential reduction in the carbon output, indicating progress toward environmental sustainability. The standard deviation of 1060.548 suggests significant variation in GHG across observations. The variable exhibits a right skewness of 6.5665 and a kurtosis of 51.7230, suggesting heavy tails with notable outliers (Table 6.2).

Regarding the level of CO_2 emissions per capita, the highest level is also indexed in China, at a value of 1.15E+10, while the lowest level is scored in Comoros (29312). The result of the mean value of CARBON is at a level of 174 million, and the skewness of 9.4187 is very high, indicating a right-skewed distribution, with an oscillating trajectory over the span of years. The standard deviation of 779 million is over four times the mean value, showing extreme variability. The value for the kurtosis of 104.4269 points to a distribution with significant tails. In terms of the electricity generated by each country, as the previous results, one of the biggest energy producers and at the same time one of the biggest polluters remains China, with a value for the EG of 8839.13. The least powerful in terms of generating electricity is Kiribati, a less developed country, while the average value for EG is at a value of 119.0697. The standard deviation of EG deviates on average from the mean by 512.6224, showing a great amount of variability. The distribution is right-skewed, as suggested by the value for the skewness of 9.6512, and the presence of outliers pointed by the sharp peak is due to the high value for the kurtosis, of 114.8354.

Table 6.2 Summary statistics

Variable	Obs	Mean	Std. Dev.	Min	Max	Skewness	Kurtosis
Y/L	3854	19034.27	21007.98	628.6933	157602.5	1.9668	7.9645
HDI	3865	0.6891	0.1616	0.2620	0.962	−0.3647	2.1702
GHG	2029	329.8029	1060.548	0.55	11,314	6.5665	51.723
CARBON	3974	1.74×10^8	7.79×10^8	29,312	1.15×10^{10}	9.4187	104.4269
EG	4103	119.0697	512.6224	0.01	8839.13	9.6512	114.8354
EC	4074	821.6437	3180.455	0.1241	44275.91	8.4161	84.1093
ECC	4072	25880.06	34841.69	105.1101	262585.7	2.6001	11.1946

Note: Authors' processing in Stata

Another category of variables included is the primary energy consumption, in which China is repeatedly a key global player and demonstrates a substantial EC of 44.275.91, while the lowest value is attributed to Kiribati (0.1240929). The mean value for this indicator is 821.6437, and the skewness of 8.4161 confirms a right-skewed distribution, with most values clustered at the lower end. The standard deviation of 3180.455 expresses considerable variability, while the kurtosis of 84.1093 highlights heavy tails with outliers. When it comes to the quantity of primary energy consumed per individual, Qatar is in the first place in the specific timeframe considered, recording a value of 262585.7. The smallest amount of ECC is 105.1101, and it belongs to each individual from Chad. Also, the mean value for ECC is 25880.06, and the skewness of 2.6001 indicates a distribution skewed to the right. The standard deviation of 34851.69 is larger than the mean value, showing variability, and the kurtosis of 11.1946 suggests moderately heavy tails. This variable reflects a more balanced distribution than the total energy consumption EC, with fewer outliers, though still reflecting disparities.

Furthermore, another important variable for the analysis is the GDP per capita, which varies significantly, recording the highest value of $157602.5 for Macao and the smallest value of $628.69 for Mozambique. The average value across the observations stands at $19034.27, with a skewness of 1.9668, with most of the values lower, but a long tail of higher values. The standard deviation of 21007.98, being slightly greater than the mean, indicates that the data points deviate on average from the mean by a substantial amount, while the kurtosis of 7.9645 leads to heavy tails, with outliers. The values reflect economic challenges that the nations may be encountering in the specified time period. Taking into account the Human Development Index, Nigeria reports the highest value of 0.962, while Niger is the country with the lowest value of 0.262, and the mean of HDI is 0.689, highlighting various levels of growth and human well-being. The negative skewness of −0.3647 suggests a slight left skew, meaning that more observations have higher HDI values, with fewer very low values. The standard deviation of 0.1616 is relatively small, suggesting moderate variability, and the kurtosis of 2.1702 below 3 indicates a platykurtic distribution with lighter tails and fewer outliers than a normal distribution.

The Jarque-Berra test assesses the normality by combining the skewness, which measures the asymmetry of the distribution, and the kurtosis, which measures the tailedness. For our original variables, the JB test statistics were extremely large, with p-values of 0, indicating evidence against the normal distribution by rejecting at 1%, 5% and 10% the null hypothesis that states that the data follows a normal distribution. The right-skewed distributions suggested by the high values for skewness and the outliers reflected by the high kurtosis initially suggested the abnormal distributions, confirmed with the JB test. Given the significant non-normality indicated by the JB test results, we applied the log-transformation to the variables GDP per capita (Y/L), HDI, GHG, CARBON, EG, EC, and ECC, in order to linearize the relationships and mitigate the impact of outliers, which are prevalent in our dataset. The only exception from further transformations is the variable HDI, which follows a normal distribution from the initial moment of analysis of its skewness and kurtosis. After the log-transformation, we reevaluated the normality of the transformed variables. The JB test statistics decreased substantially, for instance. From 6443 to 140.2 for GDP per capita and from 1.2×10^6 to 28.41 for EC, indicating that the distributions became less skewed and had reduced the kurtosis. However, the p-values remained extremely small, for example. $3.6 \times e^{-31}$ for LogGDP per capita, $6.8 \times e^{-07}$ for LogEC, suggesting that the log-transformed variables still did not achieve a perfect normal distribution. This persistent non-normality may be due to the large sample sizes, which make the JB test sensitive to even minor deviations from normality. Despite the lack of perfect normality in the log-transformed variable, the significant reduction in JB test statistics indicates that the transformation improved the distributions, making the data more suitable for the econometric modeling (Tables 6.3 and 6.4).

The correlation matrix shows how closely related different types of variables are to each other, and it helps us to understand if there exists a strong or weak relationship between pairs of proxies. A positive correlation indicates they move concurrently, whereas a negative correlation suggests they move divergently. The

Table 6.3 JB statistic

Variable	JB statistic	*P*-value
GDP per capita (Y/L)	6443	0
HDI	196.6	$2.1 \times e^{-43}$
GHG	$2.2 \times e^{+05}$	0
CARBON	$1.8 \times e^{+06}$	0
EG	$2.2 \times e^{+06}$	0
EC	$1.2 \times e^{+06}$	0
ECC	$1.6 \times e^{+04}$	0
LogGDP per capita	140.2	$3.6 \times e^{-31}$
LogGHG	15.34	$4.7 \times e^{-04}$
LogCARBON	38.95	$3.5 \times e^{-09}$
LogEG	54.91	$1.2 \times e^{-12}$
LogEC	28.41	$6.8 \times e^{-07}$
LogECC	198.9	$6.3 \times e^{-44}$

Note: Authors' processing in Stata

Table 6.4 Correlation matrix

n = 1845	GDP per capita	HDI	GHG	CARBON	EG	EC	ECC
GDP per capita	1						
HDI	0.8794	1					
GHG	0.4835	0.566	1				
CARBON	0.5168	0.6002	0.9713	1			
EG	0.5386	0.6213	0.9082	0.9141	1		
EC	0.5516	0.6242	0.9681	0.9774	0.9556	1	
ECC	0.9213	0.8475	0.5306	0.5706	0.5709	0.6027	1

Note: All the variables are log-transformed. Authors' processing in Stata

correlation matrix gives us valuable insights into how proxies interact with each other. The provided correlation matrix reveals significant relationships between economic and environmental indicators. There exists an exceedingly robust and affirmative correlation (0.8794) between GDP per capita and HDI, indicating that higher GDP per capita is closely associated with improved human development. Greenhouse gas emissions (GHG) show moderate to strong positive correlations with both GDP per capita (0.4835) and HDI (0.566), suggesting that economic growth and higher human development levels are accompanied by increased greenhouse gas emissions.

CO_2 emissions per capita (CARBON) exhibit strong correlations with HDI (0.6002) and GDP per capita (0.5168), indicating that as economies are prospering and human development is improving, carbon emissions per capita are also rising. The very high correlations of GHG with CARBON (0.9713) and EG (0.9082) demonstrate that greenhouse gas emissions are strongly linked to CO_2 emissions per capita and electricity generation. This trend is further supported by the strong relationships between EC and both GHG (0.9681) and CARBON (0.9774), indicating that primary energy consumption is a major driver of emissions. Electricity generation (EG) also shows strong correlations with HDI (0.6213) and GDP per capita (0.5386), emphasizing the function of electricity in fostering economic expansion and development. The close relationships between EG and both GHG (0.9082) and EC (0.9556) highlight the environmental impact of electricity production. Primary energy consumption per capita (ECC) has a very strong correlation with GDP per capita (0.9213) and HDI (0.8475), underscoring that as economies grow and human development improves, the energy consumption per person increases significantly. The moderate correlations of ECC with GHG (0.5306), CARBON (0.5706), and EG (0.5709) suggest that while primary energy consumption per capita is related to emissions and electricity generation, the relationship is not as strong as with total primary energy consumption.

Overall, the matrix of correlation highlights the intricate connections between economic growth and environmental impact. Economic growth, measured by GDP per capita, is strongly connected to an upward trend in energy consumption and emissions, reflecting the environmental challenges posed by economic activity. This underscores the need for sustainable development strategies that balance economic growth with environmental preservation.

Model Specification

In our fixed effects, modeling we employed an investigation of the relationship between GDP per capita, represented by Y, modeled as a function of capital (K) and labor (L). We analyze the productivity per worker by the division of the equation by labor (L), resulting in $Y/L = (K/L)$, which will be further augmented with energy (E) and control variables (C). All these variables are specific to country i in year t. The simple version of our model is specified by the equation:

$$Y_{it} = \beta_0 + \beta_1 \mathrm{HDI}_{it} + \varepsilon_{it} \tag{6.1}$$

with specific components that extend *Y* by considering, GDP per capita as dependent variable, visualizing the impact of the composite index (HDI).

In the framework of Eq. 6.1, *Y* becomes GDP per capita, providing a more comprehensive understanding of the factors driving economic output per individual. The dependent variable GDP per capita represents the gross domestic product per capita, having β_0 as the constant term. The linear parameter β_1 is associated with the independent variable HDI. The residual term, ε_{it} captures the model's error.

Moreover, in analyzing the relationship between economic growth, energy usage, and environmental quality, we adapted the simple regression model from Eq. 6.1 with an extended version, using energy consumption variables (*E*) and control variables measuring the environmental quality (EQ). This yields the extended model: $Y/L = f\,(K/L, E, \mathrm{EQ})$. Here, *K*/*L* is substituted by a composite index such as the Human Development Index (HDI), which reflects broader aspects of productivity and economic structure. We consider *E*, a vector representing three key energy variables: EC (primary energy consumption), ECC (primary energy consumption per capita), and EG (electricity generation). These proxies provide a comprehensive view of a nation's energy profile, encompassing both total energy use and its distribution across the population, as well as the output of electricity. On the environmental side, EQ represents two indicators of environmental quality: GHG (greenhouse gas emissions) and CARBON (CO_2 emissions per capita). Our regression analysis provides perspectives on the interplay between economic growth and environmental factors. The coefficients indicate how changes in HDI, GHG emissions, CO_2 emissions per capita, primary energy consumption, and electricity generation are associated with variations in GDP per capita.

The extended version using the fixed effects model is specified as follows:

$$Y_{it} = \beta_0 + \beta_1 \mathrm{HDI}_{it} + \beta_2 E_{it} + \beta_3 EQ_{it} + \varepsilon_{it} \tag{6.2}$$

The description of variables corresponds to the ones presented above. Still, Eq. 6.2 validates a(n) (inverted) U-shaped relationship between environmental quality proxies and economic growth, justified by approaching the Environmental Kuznets Curve (EKC) model, allowing us to capture non-linear dynamics, where the environmental quality has the tendency to deteriorate initially with economic growth, but then improves after reaching a certain level.

6.4 Results and Discussions

Descriptive Results

In our analysis of the parametric panel data model, the GHG variable is log-transformed and illustrated in a quadratic manner, estimating a parametric regression. This pattern exhibits a concave representation in relation to the economic growth of the nations, as a function of various environmental proxies. On the graphical representation, the highest peak is visible at a proxy threshold of 7.3175, and afterward, even if the values of the emissions are increasing, they cannot overcome the maximum, and the trend will go slightly downward. The 1% significance of the estimated coefficients suggests robust evidence that the GHG emissions have a meaningful impact on the development of countries. Also, the peak value of ECC is highly unlikely to be attained, as can be observed on the graph with a direct trend. Being significant at 1%, the CARBON decreases the expected GDP by 0.01%, the EG by 0.009%, and the EC also by 0.009%. While nations like Switzerland and Norway maintain HDI at a higher level and the emissions at a lower level, countries like China and India are challenged to balance the economic growth with a healthy and sustainable environment, having a major impact on their prosperity (Fig. 6.2).

In modeling the relationship between GDP per capita of nations worldwide as the dependent variable, with emissions, energy consumption, and production as independent variables, the lfit and qfit graphs in Stata reveal insightful trends. The lfit graph, which overlays a linear regression line on the scatterplot, demonstrates a clear positive linear relationship, indicating GDP per capita has the tendency to increase with higher levels of emissions, energy usage, and generation. This generally suggests that countries with higher emissions and energy usage also have higher development indices and economic output. Meanwhile, the qfit graph, which adds a quadratic regression approach, captures the same positive trend but with a nuanced curve. The quadratic fit suggests that the rate of increase in GDP per capita may accelerate or decelerate at higher levels of emissions and energy consumption, reflecting potential diminishing returns or escalating impacts at extreme values. These visualizations underline the complexity of the economic growth nexus, highlighting that while economic development generally rises with increased emissions and energy consumption and production, the relationship may not be strictly linear, especially at the extremes.

Estimation Results

In the results of the simple regression modeled by the fixed effects as provided by Table 6.5, positive relationship between Y is observed (GDP per capita) and K/L (HDI), indicated by the coefficient of K/L, highly significant at the 0.01 level (***). Specifically, one unit increase in K/L multiplies Y by $(e)^{4.278} = 1.3435$ and increases

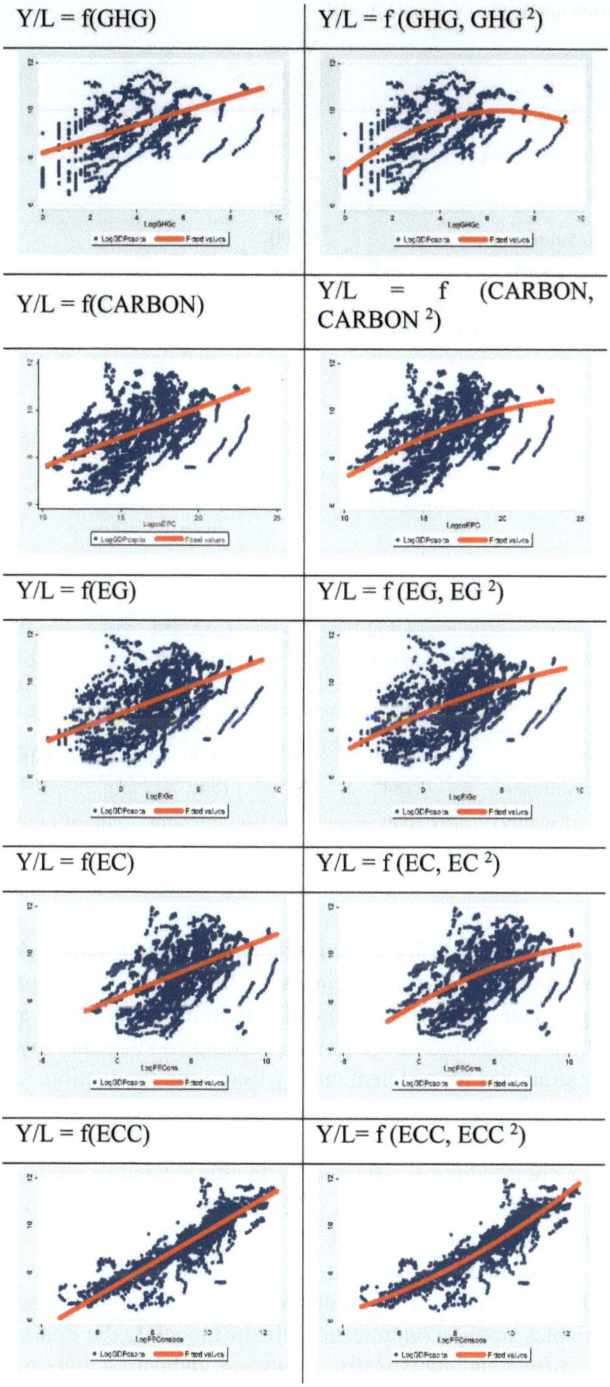

Fig. 6.2 Plots of GHG emissions against GDP per capita. (Note: All the variables are log-transformed. Authors' processing in Stata)

Table 6.5 Estimation results of simple model

Y	Model 1
K/L	4.278***
Intercept	6.264***
Within R^2	0.636
Between R^2	0.814
Overall R^2	0.806
Hausman test (*p*-value)	0.000
Number of observations	3738

EQ (environmental quality) is proxied by greenhouse gas emissions. Authors' processing in STATA
Notes: *** Significant at the 0.01 level

Table 6.6 Estimation results of extended model

Y	Model 1	Model 2	Model 3
K/L	3.508***	3.804***	4.126***
EC	0.271***		
ECC		−0.031	
EG			−0.072***
GHG	−0.03*	0.111***	0.129***
Intercept	4.472***	6.427***	6.198***
Within R^2	0.650	0.612	0.619
Between R^2	0.875	0.741	0.75
Overall R^2	0.875	0.738	0.747
Hausman test (*p*-value)	0.001	0.000	0.000
Number of observations	1845	1845	1845

EQ (environmental quality) is proxied by greenhouse gas emissions. Authors' processing in STATA
Notes: *** Significant at the 0.01 level; * Significant at the 0.10 level

the Y with 34.35%, which reflects the fundamental role of human development in boosting the productivity of a country in a specified period, further providing better access to energy usage. This tendency positively impacts developed and developing nations and drives economic development, leading to sustainability as a lifestyle and to broader strategies for efficient energy usage and production.

Tables 6.6 and 6.7 illustrated below provide the results based on a series of fixed effects models, highlighting the level of economic growth of a developed or developing country as being directly affected by factors related to energy consumption and production and thereby emissions. The division of the table arranges the main outcomes of three fixed effects models.

The results of our fixed effects models from Table 6.6, having as the dependent variable the GDP per capita (Y), reflect positive and negative impacts of the environmental variables upon economic growth. In model 1, the coefficient for HDI (K/L) is both positive and statistically significant, indicating that one unit increase

Table 6.7 Robustness checks

Y	Model 1	Model 2	Model 3
K/L	3.849***	3.977***	4.167***
EC	0.244***	–	–
ECC	–	−0.027*	–
EG	–	–	−0.063***
CARBON	−0.055***	0.074***	0.094***
Intercept	5.223***	5.358***	4.923***
Within R^2	0.668	0.637	0.644
Between R^2	0.851	0.805	0.810
Overall R^2	0.848	0.800	0.805
Hausman test (*p*-value)	0.000	0.000	0.000
Number of observations	3705	3705	3712

EQ (environmental quality) is proxied by carbon dioxide emissions per capita. Authors' processing in STATA

Notes: *** Significant at the 0.01 level; * Significant at the 0.10 level

in the HDI (K/L) corresponds to 1.52 units increase in the GDP per capita. A 1% increase in our first control variable, GHG, multiplies GDP per capita (Y) by $(e)^{-.03*\log(1.01)}$. Hence, a 1% increase in GHG will basically decrease GDP per capita by 0.01%, revealing the indirect impact of emissions on the economic growth of a nation, as it was expected. The model explains 65% of the Within R^2 and 87.5% of the Between R^2, with an overall R^2 of 87.5%. Taking a look at the graph, there is a direct trajectory up to the vertex of the parabola (6.6487), followed by an indirect impact. Basically, we may notice that the great majority of observations are located on the left side of the vertex, with GHG values in between 0 and 6.6487, corresponding to unlogged GHG values in between $e^0 = 1$ and $e^{6.6487} = 771.7803$. In this part of the parabola, as GHG increases, so does the economic growth reflected by the GDP per capita. The steepness of the curve is more abrupt for low GHG values, slightly adjusting its steepness as the values of GHG approach the vertex ones. Beyond the vertex, any increase in GHG is related to a decrease in GDP per capita, and there are fewer observations on this side of the parabola, without inferring causality. Noticing the direct impact of HDI and GHG from Model 2, the EC is not significant, suggesting that this proxy has no impact on Y in this model. The negative coefficient for EG (−0.072) indicates that increased electricity generation lowers Y/L, possibly due to some related environmental causes or inefficiencies. The models were diagnosed with the Hausman Test, which confirms the significance and appropriateness of the fixed effects model over the random effects model.

In Table 6.7, analyzing Model 1 points out that each one-unit increase of ECC after multiplying it by e ≈ 2.72, will increase the level of expected GDP per capita by 5.35%, explaining that the amount of primary energy consumption per capita will have a positive impact on the economic situation of each specific country. The ECC has the tendency to go up with the GDP per capita (Y), developed countries like the USA consuming more energy due to industrialization, while developing countries like India increase the energy usage as they industrialize. The direct influence of the CARBON on the GDP per capita reveals a significance threshold of 1%. For instance, China and Japan are among the highest polluting countries and at the same time, the most developed with an incredibly numerous populations, so our results resonate perfectly with the positive coefficients of CARBON from Model 2 and Model 3, where the Y awaits an expected increase. Model 2 exhibits a downward trend for EC, which slightly decreases Y by 0.01%, after multiplying Y with $(e)^{-.027 * \log(1.01)}$. Multiplying EG by e ≈ 2.72 multiplies Y by $(e)^{-.063} = 0.9389$, which decreases the expected GDPcapita (Y) by 6.11% in Model 3. Regarding the less developed and developing countries, the increasing trend of economic growth is higher in comparison to the countries that are well developed, since their rise is linear and steady.

To validate the consistency of our main results, we performed a robustness check considering an alternative measure by adding some alternative control proxies for environmental quality, namely CARBON (CO_2 emissions) and GHG (greenhouse gas emissions), to ensure the validity of our results across different models.

In all models, HDI (K/L) consistently demonstrates a strong and positive correlation with economic output per worker, highlighting the crucial role of human development. The energy variables have mixed effects; ECC positively affects Y, EC has a negative outcome, whereas EG also has a negative impact, possibly due to inefficiencies in production. The environmental quality variable (GHG) reveals a complex pattern, transitioning from a negative to a positive influence on Y as additional energy variables are included, suggesting that emissions may increase productivity while at the same time being a threat to a sustainable life. The Hausman test consistently validates the robustness of the fixed effects model across all the models. Moreover, the connection between economic growth and environmental quality suggests that economic and human well-being may influence economic growth, subsequently impacting GDP per capita.

6.5 Conclusion

The rapid increase in urban development, energy usage, and emissions worldwide has driven the need to explore the connections between these factors. To achieve this, a panel model was applied to data spanning from 2000 to 2022, covering 185 countries globally. The empirical results of this study reveal direct and indirect connections between human development, GHG emissions, CO_2 emissions, energy usage, electricity generation, on the one hand, and GDP per capita on the other hand. The independent proxies are found to have a significant impact on the development from an economic standpoint and a sustainable perception, in both developed and developing countries as provided by our sample. However, the impact varies by nation, and it is quite meaningful regarding the developing countries. In less developed and developing countries, prosperity tends to increase at a faster rate than in well-developed countries, where growth is more gradual and consistent.

Many of these countries are experiencing rapid economic growth, which is correlated with a negative outcome regarding the rise of polluting emissions and energy usage and generation. For instance, the title for one of the biggest countries in terms of population and economic structure belongs to China, being in a constant hurry on its economic growth path, yet not reaching decreases in the context of energy and emissions, but accelerating them. Salahuddin, Gow, and Ozturk (2015), Agusalim et al. (2025), and Lu, Ma, and Hu (2024) highlight the vital connection between energy consumption and economic growth, predicting that countries will face international pressure to reduce polluting emissions and the risk of energy shortages as the path of urbanization continues to expand rapidly. To successfully mitigate the negative implications of energy consumption and emissions upon prosperous outcomes in terms of sustainability, governments should be more supportive, especially in the developing and less developed countries, in order to attain improved effects and be the change that our planet is craving for to reduce the challenges posed by climate change. In this manner, promoting economic growth within a cleaner environment will also be helpful in bridging the gap between nations.

Efforts should be made to accelerate the phasing out of obsolete production technologies and equipment while fostering technological advancement in the industrial sector. In discussing the impact of polluting emissions and energy usage upon the economic prosperity and human development of nations, it is essential to consider both developed and developing countries. Zaman and Moemen (2017) found that in 90 countries, the relation of economic growth with CO_2 emissions forms an inverted-U shape. However, in high-income nations such as Australia, the USA, and the United Kingdom, this relationship levels off. This suggests that these wealthy countries have reached a point in their economic growth where further growth doesn't significantly increase pollution and this can be attributed to advanced technology and innovation in key economic sectors in these high-income countries, unlike in developing and least developed countries. Developing nations should leverage their abundant natural resources in order to foster renewable energy production. By doing so, they can mitigate environmental pollution and stimulate economic prosperity. On the other hand, developed nations should reduce their dependency on fossil fuel imports, which would not only decrease environmental degradation but also strengthen the upward trend of economic growth.

Policy recommendations should thus include collaborative efforts between developed and developing nations to utilize advanced technologies for the sustainable extraction of resources such as fuel. This is especially crucial for developing countries, which should also focus on increasing their fuel exports, as these exports imply a significant contribution to reducing environmental pollution. Additionally, policymakers should craft strategies that encourage foreign investments in clean energy projects. In this manner, they ensure that international investments align with sustainable development goals, promoting clean energy infrastructure and diminishing the global carbon footprint. The analysis of GDP per capita against emissions, energy consumption, and production, despite showing positive trends, is constrained by missing observations and unvalidated variables. These gaps can introduce bias and limit the reliability of our findings. Future research should address these limitations by incorporating more control variables to account for other influencing factors. Employing methodologies such as dynamic panel modeling can better handle unobserved heterogeneity, while quantile regression could provide insights across different distribution points of GDP per capita. The use of dummy variables can also help in capturing categorical influences, such as regional effects or policy differences, enhancing the robustness and depth of the analysis. These advancements would yield a more comprehensive understanding of the development-environment dynamics.

In summary, an integrated approach that includes utilizing natural resources for renewable energy, reducing the dependence on fuel imports, employing advanced technologies for sustainable resource extraction, increasing the fuel exports, and fostering international investments in clean energy can collectively enhance both economic growth and human development while addressing environmental concerns.

References

Agusalim, L., Karim, M., Irfany, M. I., Lestari, A. V. A., & Pebrianti, H. (2025). Economic development and climate change: Insights from religious countries. *International Journal of Energy Economics and Policy, 15*(2), 58–83. https://doi.org/10.32479/ijeep.17975

Akhmat, G., Zaman, K., Shukui, T., et al. (2014). Does energy consumption contribute to environmental pollutants? Evidence from SAARC countries. *Environmental Science and Pollution Research, 21*(9), 5940–5951. https://doi.org/10.1007/s11356-014-2528-1

Eyuboglu, K., & Uzar, U. (2022). Asymmetric causality between renewable energy consumption and economic growth: Fresh evidence from some emerging countries. *Environmental Science and Pollution Research, 29*, 21899–21911. https://doi.org/10.1007/s11356-021-17472-9

Gershon, O., Asafo, J. K., Nyarko-Asomani, A., & Koranteng, E. F. (2024). Investigating the nexus of energy consumption, economic growth and carbon emissions in selected African countries. *Energy Strategy Reviews, 51*, 101269. https://doi.org/10.1016/j.esr.2023.101269

Haseeb, A., Xia, E., Danish, Baloch, M. A., & Abbas, K. (2018). Financial development, globalization, and CO2 emission in the presence of EKC: evidence from BRICS countries. *Environmental science and pollution research international, 25*(31), 31283–31296. https://doi.org/10.1007/s11356-018-3034-7

Intergovernment al Panel on Climate Change (IPCC). (2022). Summary for policymakers. In *Global warming of 1.5°C: IPCC special report on impacts of global warming of 1.5°C above pre-industrial levels in the context of strengthening the response to climate change, sustainable development, and efforts to eradicate poverty* (pp. 1–24). Cambridge University Press.

International Energy Agency (IEA). (2023). *CO_2 emissions in 2022*. https://www.iea.org/reports/co2-emissions-in-2022. Accessed 20 Jan 2025.

Kulu, E., Karakara, A. K., Dasmani, I., & Armah, M. K. (2024). Energy consumption-economic growth nexus: How do sectoral differentials affect sustainable development? *Sustainable Development*. https://doi.org/10.1111/1477-8947.12390

Lee, C.-C., & Chang, C.-P. (2007). Energy consumption and GDP revisited: A panel analysis of developed and developing countries. *Energy Economics, 29*(6), 1206–1223. https://doi.org/10.1016/j.eneco.2007.01.001

Lefatsa, P., & Nubong, G. (2025). Financial development, economic growth, and energy consumption in SADC region. *BRICS Journal of Economics, 6*(1), 223–258. https://doi.org/10.3897/brics-econ.6.e138454

Lu, F., Ma, F., & Hu, S. (2024). Does energy consumption play a key role? Re-evaluating the energy consumption-economic growth nexus from GDP growth rates forecasting. *Energy Economics, 129*, 107268. https://doi.org/10.1016/j.eneco.2023.107268

Magazzino, C., Mele, M., Schneider, N., & Vallet, J. (2021). Energy consumption and economic growth in Italy: A wavelet analysis. *Energy Reports, 7*, 1520–1528. https://doi.org/10.1016/j.egyr.2021.03.005

Patlins A, Caiko J, Kunicina N, Zhiravetska A, Riashchenko V (2020) Climate education: challenges of climate change and energy policies. In: 2020 IEEE 61st International Scientific Conference on Power and Electrical Engineering of Riga Technical University (RTUCON). IEEE, pp 1–5. https://doi.org/10.1109/RTUCON51174.2020.9316561

Riebeek H (2010) Global warming. NASA Earth Observatory. https://science.nasa.gov/earth/climate-change/global-warming/. Accessed 17 Dec 2025

Ritchie, H., & Roser, M. (2020). *CO_2 emissions. Our world in data*. https://ourworldindata.org/co2-emissions. Accessed 25 Feb 2025.

Salahuddin in, M., Gow, J., & Ozturk, I. (2015). Is the long-run relationship between economic growth, electricity consumption, carbon dioxide emissions and financial development in Gulf Cooperation Council Countries robust? *Renewable and Sustainable Energy Reviews, 44*, 314–325. https://doi.org/10.1016/j.rser.2015.06.005

Salari, M., Kelly, I., Doytch, N., & Javid, R. J. (2021). Economic growth and renewable and non-renewable energy consumption: Evidence from the U.S. states. *Renewable Energy, 178*, 50–65. https://doi.org/10.1016/j.renene.2021.06.016

Tran, B.-L., Chen, C.-C., & Tseng, W.-C. (2022). Causality between energy consumption and economic growth in the presence of GDP threshold effect: Evidence from OECD countries. *Energy, 251*, 123902. https://doi.org/10.1016/j.energy.2022.123902

Triantafyllidou, A., & Polychronidou, P. (2025). Energy consumption, economic growth and CO_2 emissions: Empirical evidence for EU countries. *Scientific Annals of Economics and Business, 72*(1), 59–78.

United Nations (2017) The Sustainable Development Goals Report 2017. https://unstats.un.org/sdgs/report/2017/. Accessed 17 Dec 2025

United Nations Environment Programme (UNEP) (2017) The Emissions Gap Report 2017: a UN Environment synthesis report. https://www.unep.org/resources/emissions-gap-report-2017. Accessed 17 Dec 2025

United Nations Framework Convention on Climate Change (UNFCCC) (2015) The Paris Agreement. https://unfccc.int/process-and-meetings/the-paris-agreement. Accessed 17 Dec 2025

Wei, T., Dong, W., Yan, Q., Chou, J., Yang, Z., & Tian, D. (2015). Developed and developing world contributions to climate system change based on carbon dioxide, methane and nitrous oxide emissions. *Advances in Atmospheric Sciences*. https://doi.org/10.1007/s00376-015-5141-4

Zaman, K., & Moemen, M. (2017). Energy consumption, carbon dioxide emissions and economic development: Evaluating alternative and plausible environmental hypotheses for sustainable growth. *Renewable and Sustainable Energy Reviews, 74*, 1119–1130. https://doi.org/10.1016/j.rser.2017.02.072

Zhao, D., Yuan, J., Fu, S., Song, Y., Wang, Y., Liu, Y., & Zhang, J. (2023). Does economic growth stimulate energy consumption? New evidence from national and regional levels in China. *Chinese Journal of Population, Resources and Environment, 21*(2), 60–70. https://doi.org/10.1016/j.cjpre.2023.06.003

Chapter 7
Leadership Structure and Implications of Partial State Ownership in the Hospitality Sector

Flávio Ferreira and Jorge Umbelino

Abstract Privatization frequently boosts efficiency and productivity in businesses as private companies are generally more motivated to reduce costs, innovate, and meet market needs than state-run counterparts. The privatization of the hospitality sector pertains to the acquisition or operational control of hospitality properties, such as hotels and resorts, including additional services, shifting from government provision to private companies. This shift can have mixed effects on environmental sustainability.

This paper delves into the interplay of environmental taxes, pollution control spending, and the privatization of a partially state-owned (PSO) hotel within a mixed duopoly framework. The market we analyze comprises a single partially state-owned (PSO) hotel competing against one for-profit (FP) hotel in a five-stage game: (i) the government decides how much of the PSO hotel will be privatized; (ii) a regulator, aiming to maximize social welfare, establishes the emission tax rate; (iii) the PSO hotel selects abatement pollution investments; (iv) the FP hotel then selects pollution abatement investments; and (v) subsequently, the two hotels concurrently and autonomously determine the quantity of rooms available for reservation. This game presents a model of a policy regime featuring commitment. In contrast, our analysis also considers a non-committed regime, distinguished by the fact that step (ii) occurs after decisions regarding abatement pollution investments.

This study's most significant finding is that, according to the social welfare perspective within the analyzed models, neither total privatization nor complete nationalization represents the optimal governmental strategy.

Keywords Partial privatization · Abatement pollution · Environmental policy

---

F. Ferreira (✉)
UNIAG, CiTUR, ESHT, Polytechnic of Porto, Vila do Conde, Portugal
e-mail: fff@esht.ipp.pt

J. Umbelino
Estoril Higher Institute for Tourism & Hotel Studies (Portugal), CiTUR, Centre for Tourism Research, Development, and Innovation, Estoril, Portugal
e-mail: jorge.umbelino@eshte.pt

A. L. Negrusa, M. M. Coroş (eds.), *Innovation, Sustainability, and Growth in a VUCA Environment*, Springer Proceedings in Business and Economics,
https://doi.org/10.1007/978-3-032-11639-0_7

7.1 Introduction

Governments in many countries strategically employ privatization to boost economic efficiency, raise revenue, elevate service quality, and spur economic growth. These privatization efforts span diverse economic sectors where governments might own and manage enterprises, including areas like airports, banking, energy, telecommunications, hospitals, and the hospitality industry.

The Angolan government, under the authority of President João Lourenço, initiated a public tender at the start of 2024 for the privatization of 39 hotels from the IKA, IU, and BINA groups, which are distributed across the nation's 18 provinces (Further Africa, 2024). On April 11, 2025, the Angolan Government announced that it will privatize eight hotel units of the IU and Bina chains, through a limited tender by prior qualification, under the Privatization Program (ProPriv) 2023–2026 (Jornal de Negócios, 2025).

In late 2023, ICON, a subsidiary of Talaat Mostafa Holding Group, took over a majority stake in seven hotels. This was a key part of Egypt's ongoing privatization efforts (Egypt Today, 2023).

As highlighted in a 2015 report by UN Tourism (known as UNWTO until 2023), Public-Private Partnerships can lead to inclusive, sustainable, and resilient tourism development, offering a pathway to contribute to the new Sustainable Development Goals. Regarding the measurement of sustainable performance in the hospitality sector, Franzoni et al. (2021) emphasize environmental aspects, outlining indicators such as CO_2 emissions, energy use per guest night, renewable energy production, water use, plastic consumption, environmental certifications, and whether certified eco-friendly cleaning products are utilized. This paper also focuses on environmental sustainability, highlighting the crucial role of incorporating the "3Rs" (reduce, reuse, and recycle) into hospitality companies' environmental strategies. Chen and Peng (2016) and Ioannidis et al. (2021) are among the researchers who highlight how crucial these principles are for boosting overall performance and success.

Driven by worsening environmental degradation and increased public ecological awareness, governments are increasingly prioritizing environmental concerns. Research has explored the link between privatization and environmental outcomes. Xing et al. (2022) analyze the impact of partial privatization on environmental R&D for clean technology and its effects on the environment and social welfare in a polluting mixed duopoly. Studies by Haruna and Goel (2019), Ferreira and Ferreira (2013), and Wang and Wang (2009) examine the effects of complete privatization on endogenously determined environmental taxes, while Ouattara (2019), Xu et al. (2016), and Pal and Saha (2015) examine partial privatization's effects on similar topics. In contrast, Xing et al. (2020) examine exogenous environmental taxes when studying the impact of complete privatization on environmental R&D and the environment. Finally, Dong et al. (2019) studied the government's simultaneous decisions on public firm privatization and the choice between environmental taxes and emission standards.

Leal et al. (2018) investigated how the strategic selection of abatement technology interacts with the government's environmental policy commitment timing in a market with both consumer-oriented and profit-driven firms, considering both committed and non-committed scenarios. Expanding on this, Ferreira et al. (2022) analyzed sequential decisions in pollution abatement investments under both policy regimes. Their research showed that when consumer-friendliness is high, a government's committed environmental tax is lower than if it doesn't commit. Interestingly, they also found that while a non-committed policy might result in less environmental damage, social welfare is consistently improved when the government makes a firm commitment to its environmental policy.

This paper unfolds as follows: We begin by developing the model in the following section. Subsequently, we investigate committed and non-committed policy regimes, report on what we find, and compare them. The paper concludes with a discussion of our findings.

7.2 Research Purpose and the Methodological Approach

The present study advances the existing literature by developing a model that aligns with the work of Leal et al. (2018), while incorporating the novel aspects of partial state ownership and sequential decision-making, tailored to the dynamics of the hospitality sector. We observe that the hospitality sector, particularly hotels and resorts, is a major contributor to national and global economies. It generates significant revenue, creates numerous jobs, and plays a crucial role in tourism, which often drives other related industries. Studying competition within such a vital sector provides insights into its efficiency, growth, and stability. Employing a mixed duopoly model, our analysis examines the interplay among environmental taxation, pollution abatement investments, and the privatization of a hotel with partial state equity. In contrast to Ferreira and Ferreira (2025), who modeled a similar problem with simultaneous decisions regarding pollution abatement investments, our work examines a sequential decision-making process.

To investigate our research question, we will leverage a game-theoretical model to analyze the decision-making process. Game theory, a mathematical framework, allows us to study interdependent choices made by multiple players. This approach helps us anticipate the strategies these players might employ to optimize their outcomes, anticipating others' moves. Frequently employed in industrial organization studies, game theory provides valuable insights into competitive and cooperative behaviors within contexts like market competition.

To manage the intricate nature of the research, the study focuses solely on two hotels. Distinguishing the two is their ownership structure: one is a partially state-owned (PSO) hotel, while the other prioritizes profit maximization (PM).

The representative consumer's maximization problem is given by

$$U(q_1,q_2) - p_1q_1 - p_2q_2$$

with $U(q_1,q_2) = q_1 + q_2 - \frac{1}{2}(q_1 + q_2)^2$ and where q_i denotes the quantity of hotel H_i rooms offered for rent to guests, p_i signifies the associated room rate, with $i = 1, 2$. Therefore, direct demand can be defined by

$$p_i = 1 - q_i - q_j,$$

where $i, j = 1, 2$ with $i \neq j$. Both hotels share the same cost function, which is given by $C_i = q_i^2 / 2$. Hotel operations cause environmental pollution $e_i > 0$. Each hotel, though, has the capacity to diminish this pollution down to a level $e_i = q_i - a_i$ by allocating an investment of $a_i^2 / 2$ toward pollution abatement initiatives. The aggregate environmental damage is quantified as $ED = \frac{(e_1 + e_2)^2}{2}$, while the marginal environmental damage is represented by $MED = e_1 + e_2$. To address pollution, the government fixes a tax of t on every unit of emissions, resulting in a total tax revenue of $T = t(e_1 + e_2)$. Therefore, the profit function π_i for hotel H_i is given by

$$\pi_i = p_iq_i - \frac{1}{2}q_i^2 - te_i - \frac{1}{2}a_i^2. \tag{7.1}$$

Social welfare W is determined by the relationship

$$W = CS + \pi_1 + \pi_2 + T - ED, \tag{7.2}$$

with $CS = \frac{1}{2}(q_1 + q_2)^2$ being the consumer surplus. Ownership of the PSO hotel is partially held by the government, with a share of $(1 - \alpha) \in [0, 1]$. The PSO hotel H_1 aims to maximize a weighted average of social welfare and its own profit, with its objective function given by

$$V = \alpha\pi_1 + (1 - \alpha)W \tag{7.3}$$

(Matsumura, 1998; Han & Ogawa, 2015). When $\alpha = 1$, the hotel operates as a fully private entity, with the manager focused on maximizing profits. Conversely, if $\alpha = 0$, the hotel is fully nationalized, and the manager's objective shifts to maximizing social welfare. Hotel H_2 operates with the objective of maximizing its profit

$$\pi_2 = p_2q_2 - \frac{1}{2}q_2^2 - te_2 - \frac{1}{2}a_2^2. \tag{7.4}$$

We analyze two policy approaches: (i) committed and (ii) non-committed. In both scenarios, the model unfolds as a five-stage game.

In the committed policy approach, the game unfolds in these steps:

- In the first stage, the government sets the level of privatization of the partially state-owned hotel.
- In the second stage, the welfare-maximizing regulator fixes the environmental tax.
- In the third stage, PSO hotel chooses abatement pollution investments.
- In the fourth stage, PM hotel sets abatement pollution investments.
- In the fifth stage, both hotels simultaneously and independently decide on the number of rooms they will offer for rent to guests.

In the non-committed policy approach, the game unfolds in these steps:

- In the first stage, the government sets the level of privatization of the partially state-owned hotel.
- In the second stage, PSO hotel chooses abatement pollution investments.
- In the third stage, PM hotel sets abatement pollution investments.
- In the fourth stage, the welfare-maximizing regulator fixes the environmental tax.
- In the fifth stage, both hotels simultaneously and independently decide on the number of rooms they will offer for rent to guests.

To find the subgame perfect Nash equilibrium, we will solve the game using backward induction. This method involves determining the equilibrium values sequentially, starting from the final stage and working backward to the first.

This study contributes to the field by exploring a market situation not yet addressed in the literature.

7.3 Results I: The Committed Policy Regime

In the fifth stage, the PSO hotel H_1 determines the output q_1 to maximize V, and the PM hotel H_2 determines the output q_2 to maximize π_2. Solving the system

$$\begin{cases} \dfrac{\partial V}{\partial q_1} = 0 \\ \dfrac{\partial \pi_2}{\partial q_2} = 0 \end{cases},$$

yields the equilibrium

$$q_1 = \frac{1+\alpha+3(1-\alpha)(p_1+p_2)+2(1-2\alpha)t}{7+\alpha}, \quad q_2 = \frac{2-(1-\alpha)(p_1+p_2)-(3-\alpha)t}{7+\alpha} \tag{7.5}$$

Now, at fourth stage, we use eqs. (7.5) into (7.3), and solve $\frac{\partial \pi_2}{\partial a_2} = 0$, which gives

$$a_2 = \frac{2\left(3(1-\alpha)-\left(29+\alpha+2\alpha^2\right)t\right)-3a_1(1-\alpha)^2}{2\left(\alpha^2-10\alpha-23\right)}. \tag{7.6}$$

At third stage, using eqs. (7.5) and (7.6) into V, we compute $\frac{\partial V}{\partial a_1} = 0$, that yields

$$a_1 = \frac{2\left((1-\alpha)\left(632+398\alpha+75\alpha^2-16\alpha^3-\alpha^4\right)-f_1(\alpha)t\right)}{4036+1557\alpha-908\alpha^2-578\alpha^3-11\alpha^5}, \tag{7.7}$$

where $f_1(\alpha) = 1494 - 1437\alpha - 1555\alpha^2 - 549\alpha^3 + 9\alpha^4 - 10\alpha^5$.

Therefore,

$$a_2 = \frac{3(1-\alpha)\left(148+41\alpha-47\alpha^2-13\alpha^3-\alpha^4\right)-2f_2(\alpha)t}{4036+1557\alpha-908\alpha^2-578\alpha^3-11\alpha^5}, \tag{7.8}$$

where $f_2(\alpha) = 2447 + 294\alpha - 568\alpha^2 - 130\alpha^3 + 9\alpha^4 - 4\alpha^5$.

At second stage, using eqs. (7.7) and (7.8) into W, we compute $\frac{\partial W}{\partial t} = 0$, that yields[1]

$$t^C = \frac{f_3(\alpha)}{f_4(\alpha)},$$

where

$$\begin{aligned} f_3(\alpha) = {} & 8346448 - 414696\alpha - 1490239\alpha^2 + 5633114\alpha^3 + 5146694\alpha^4 \\ & +1479474\alpha^5 + 232808\alpha^6 - 58098\alpha^7 + 874\alpha^8 - 1970\alpha^9 - 41\alpha^{10} \end{aligned}$$

$$\begin{aligned} f_4(\alpha) = {} & 44525152 + 2213520\alpha + 9064223\alpha^2 + 43280564\alpha^3 + 33034088\alpha^4 \\ & +10441764\alpha^5 + 1995134\alpha^6 + 88284\alpha^7 + 62128\alpha^8 - 1988\alpha^9 + 619\alpha^{10} \end{aligned}$$

As a result, the other equilibrium values now stand at

$$q_1^C = \frac{f_5(\alpha)}{f_4(\alpha)},$$

[1] Throughout the paper we use the notation superscript C to refer to the committed policy regime.

where

$$f_5(\alpha) = 6\big(2385276 - 914481\alpha + 539754\alpha^2 + 1982778\alpha^3 + 1177010\alpha^4 + 108336\alpha^5 - 24162\alpha^6 - 16490\alpha^7 + 5370\alpha^8 - 591\alpha^9 + 80\alpha^{10}\big),$$

$$q_2^C = \frac{f_6(\alpha)}{f_4(\alpha)},$$

where

$$f_6(\alpha) = 6(7+\alpha)\big(173548 + 39613\alpha + 52404\alpha^2 + 196885\alpha^3 + 137154\alpha^4 + 46377\alpha^5 + 8512\alpha^6 + 731\alpha^7 + 126\alpha^8 + 10\alpha^9\big),$$

$$a_1^C = \frac{f_7(\alpha)}{f_4(\alpha)},$$

where

$$f_7(\alpha) = 2\big(3882632 - 607482\alpha + 2455279\alpha^2 + 2902684\alpha^3 + 850570\alpha^4 + 32124\alpha^5 - 93236\alpha^6 + 10716\alpha^7 + 2822\alpha^8 + 1094\alpha^9 - 19\alpha^{10}\big),$$

$$a_2^C = \frac{f_8(\alpha)}{f_4(\alpha)},$$

where

$$f_8(\alpha) = 5222584 + 1996134\alpha - 1720477\alpha^2 + 3032456\alpha^3 + 6221852\alpha^4 + 3113460\alpha^5 + 914378\alpha^6 + 81960\alpha^7 + 11764\alpha^8 + 118\alpha^9 + 139\alpha^{10}.$$

At the initial stage, the government determines the extent of privatization for the partly state-owned hotel. Solving $\frac{\partial W}{\partial \alpha} = 0$, it results that there exists $\alpha_0^C \in (0.34, 0.35)$, such that

$$\frac{\partial W^C}{\partial \alpha} \begin{cases} > 0, \alpha \in (0, \alpha_0^C) \\ < 0, \alpha \in (\alpha_0^C, 1) \end{cases}. \tag{7.9}$$

Based on (Eq. 7.9), we derive one of the key findings of this paper.

Proposition 1. From a social welfare standpoint, the committed policy regime implies that the government should not fully privatize or fully nationalize.

Thus, considering what's best for society, and given the existing, relatively fixed rules and practices the government is operating under, the analysis suggests that the optimal level of government involvement in the (presumably partially state-owned hotel) is somewhere in between complete private ownership and complete state ownership. Neither extreme is predicted to yield the best social welfare outcomes under these specific conditions. The "committed policy regime" likely creates a context where a mix of public and private involvement, or a specific degree of partial privatization, strikes a better balance for social welfare than extreme. This could be due to factors within the regime that influence market dynamics, the distribution of benefits, or the government's ability to regulate effectively under different ownership structures.

Based on the derived equilibrium values of the decision variables, it follows that

$$\frac{\partial t^C}{\partial \alpha} < 0, \frac{\partial q_1^C}{\partial \alpha} < 0, \frac{\partial q_2^C}{\partial \alpha} < 0,$$

$$\frac{\partial ED^C}{\partial \alpha} \begin{cases} < 0, \alpha \in [0,\alpha_1) \cup (\alpha_2,1] \\ > 0, \alpha \in (\alpha_1,\alpha_2) \end{cases}, \ \alpha_1 \in (0.49,0.50), \ \alpha_2 \in (0.96,0.97).$$

Hence, we present the following result.

Proposition 2. Under the committed policy regime, the privatization level of the PSO hotel exhibits the following relationships: it negatively correlates with the environmental tax rate and the number of rooms offered by the PSO hotel, while positively correlating with the number of rooms offered by the PM hotel. Additionally, the impact of PSO hotel privatization on total environmental damage is ambiguous, potentially leading to either an increase or a decrease.

The previous finding can be interpreted as follows: a higher degree of PSO hotel privatization likely results in a lower environmental tax burden, possibly due to the operational advantages of private management or successful lobbying efforts. Simultaneously, privatizing the PSO hotel appears to cause a decrease in its available rooms, perhaps driven by profit-maximizing strategies. On the other hand, the growth in PSO hotel privatization seems to incentivize PM hotels to increase their room capacity, suggesting a competitive dynamic at play.

7.4 Results II: The Non-committed Policy Regime

In the non-committed policy regime, the fifth stage is identical to the one described above. The fourth stage involves the government's decision on the environmental tax rate. The goal is to choose the rate that maximizes social welfare W. To find this optimal rate, we substitute the previously derived results from (Eq. 7.5) into the social welfare function (Eq. 7.2), and then solve the first-order condition for maximization, $\frac{\partial W}{\partial t} = 0$. This calculation yields the optimal environmental tax rate

$$t = \frac{1-4(a_1+a_2)}{5}. \tag{7.10}$$

We see that hotels possess a strategic lever over their environmental tax burden: by proactively investing in emission reduction technologies, they can expect a reduction in their environmental tax obligations.

In the third stage, PM hotel chooses its abatement efforts. Substituting (Eqs. 7.5 and 7.10) into the profit function π_2, the first-order condition $\frac{\partial \pi_2}{\partial a_2} = 0$ leads to the abatement level

$$a_2 = \frac{8-17a_1}{62}. \tag{7.11}$$

Now, in the second stage, PSO hotel chooses its abatement efforts. Substituting (Eqs. 7.5, 7.10, and 7.11) into the objective function V, the first-order condition $\frac{\partial V}{\partial a_1} = 0$ leads to the abatement level[2]

$$a_1^{NC} = \frac{2(518-143\alpha)}{5348+2717\alpha}.$$

Thus,

$$a_2^{NC} = \frac{406+429\alpha}{5348+2717\alpha}.$$

As a result, the other equilibrium values now stand at

$$t^{NC} = \frac{3(143\alpha-28)}{5348+2717\alpha}, q_1^{NC} = q_2^{NC} = \frac{2(679+286\alpha)}{5348+2717\alpha}.$$

At the initial stage, the government determines how much of the partially state-owned hotel will be privatized. Solving $\frac{\partial W}{\partial \alpha} = 0$, we get $\alpha = 0$. So, we state the following result.

Proposition 3. From a social welfare standpoint, the non-committed policy regime implies that the government should fully nationalize.

---

[2] Throughout the paper we use the notation superscript NC to refer to the non-committed policy regime.

The preceding result suggests that if governments cannot credibly commit to future policies (taxes, regulations, subsidies), private firms may distrust them and act inefficiently. To avoid this welfare loss, full nationalization of key sectors, with the government handling ownership and investment, might be the best solution.

The analysis of the preceding result reveals that

$$\frac{\partial t^{NC}}{\partial \alpha} > 0, \frac{\partial q_i^{NC}}{\partial \alpha} < 0\left(i = 1,2\right), \frac{\partial ED^C}{\partial \alpha} > 0.$$

Hence, we present the following result.

Proposition 4. In the non-committed policy regime, greater privatization of the PSO hotel leads to a higher environmental tax rate and fewer rooms offered by both PSO and PM hotels. Moreover, as the PSO hotel becomes more privatized, total environmental damage rises.

The implication of the previous result is that privatized hotels may increase their environmental impact through their activities or expansion, potentially prompting the government to respond with higher taxes. This contrasts sharply with the committed policy regime and highlights the risk that a focus on financial performance under privatization could lead to increased environmental degradation.

7.5 Results: Comparisons

Now, we compare the outcomes we found for the committed and non-committed policy regimes.

Through straightforward calculation, we find that

$$t^C - t^{NC} > 0, ED^C - ED^{NC} < 0, W^C - W^{NC} \begin{cases} < 0, \alpha \in \left[0,\alpha_1\right) \\ > 0, \alpha \in \left(\alpha_1,1\right] \end{cases}, \text{where } \alpha_1 \in \left(0.31,0.32\right),$$

thereby producing the following results:

Proposition 5. Although the committed policy regime levies a steeper environmental tax than the non-committed one, it leads to less overall environmental harm. Interestingly, the difference in social welfare between the two regimes is not consistent; it can be positive or negative depending on the degree of PSO hotel privatization.

The previous result can be interpreted as follows. The committed regime effectively reduces environmental damage by imposing higher taxes. However, its impact on social welfare is complex and inconsistent, as it depends on external factors like the level of PSO hotel privatization. This highlights the intricate balance between environmental objectives, economic impacts, and societal well-being in policy design.

The fact that the difference in social welfare between committed and non-committed regimes varies with the degree of PSO hotel privatization is a critical insight. This suggests there's no "one-size-fits-all" answer. Policymakers can't simply assume that a committed environmental policy will always lead to greater social welfare during privatization. Instead, they must carefully consider:

- The specific level of privatization being pursued: Different levels of state ownership retention might interact differently with environmental policy commitment and affect social welfare.
- The broader economic and social context: Factors beyond environmental harm and hotel profit will influence overall social welfare.

In essence, policymakers navigating privatization in high-stakes sectors like tourism face a delicate balancing act. While a firm commitment to environmental taxation appears beneficial for the environment, they need to conduct detailed analyses specific to the degree of privatization planned to understand the full impact on social welfare. This calls for a tailored approach rather than a blanket policy, ensuring that privatization efforts genuinely serve both environmental protection and broader societal well-being.

7.6 Conclusions

This paper delved into the interconnectedness of environmental taxation, investments in pollution reduction, and the privatization of a partly government-owned hotel. Our analysis highlights the nuanced ways these elements influence both environmental quality and societal welfare. Specifically, our model posits that each hotel has the capacity to invest in lowering its pollution output, and we examined the crucial role of the government's commitment timeline for environmental tax policy on these investment decisions.

The study indicates that without firm policy commitments, increased privatization tends to lead to greater environmental damage, potentially because private entities focus on profitability at the expense of sustainability. However, the outcome under committed policies is less clear. This analysis emphasizes that policymakers must carefully consider the interplay between privatization and environmental regulation. They need to develop tax policies that not only discourage environmentally damaging behavior but also take into account the economic ramifications for businesses and society.

Balancing the advantages of privatization with the essential need for robust environmental regulations is a key challenge for policymakers. They must develop tax structures that simultaneously disincentivize environmentally damaging behavior and acknowledge the resulting economic impacts on both businesses and the community.

Acknowledgements This work was supported by national funds through FCT/MCTES (PIDDAC): UID/04752, Applied Management Research Unit (UNIAG).

References

Chen, X., & Peng, Q. (2016). A content analysis of corporate social responsibility: Perspectives from China's top 30 hotel-management companies. *Hospitality & Society, 6*(2), 153–181.

Dong, Q., Bárcena-Ruiz, J. C., & Garzón, M. B. (2019). Privatization and environmental policy in a mixed oligopoly. *Estudios de Economia, 46*(2), 173–190.

Egypt Today. (2023). https://www.egypttoday.com/Article/3/129289/ICON-acquires-majority-stake-in-7-hotels-as-part-of. Accessed 3 Mar 2025.

Ferreira, F. A., & Ferreira, F. (2013). Privatization in a mixed duopoly with environmental taxes. *AIP Conference Proceedings, 1558*, 1558–1561.

Ferreira, F., & Ferreira, F. A. (2025). Partial state ownership in the hospitality industry. In V. Katsoni & C. Costa (Eds.), *Innovation and creativity in tourism, business and social sciences* (pp. 365–376). Springer Proceedings in Business and Economics.

Ferreira, F., Ferreira, F. A., & Fernandes, P. O. (2022). Committed and non-committed policy regimes in a sequential mixed hotel duopoly. In V. Katsoni & A. C. Şerban (Eds.), *Transcending borders in tourism through innovation and cultural heritage* (pp. 91–103). Springer.

Franzoni, S., Sarwar, H., & Ishaq, M. (2021). The mediating role of HRM in the relationship between CSR and performance in the hospitality industry. *Sustainability, 13*(24), 13699.

Further Africa. (2024). https://furtherafrica.com/2024/01/17/angola-authorizes-privatization-of-39-hotels/. Accessed 3 Mar 2025.

Han, L. H. & Ogawa, H., 2015. The relationship between privatization and regulation on foreign investment policies. *Theoretical Economics Letters, 5*, 97–102.

Haruna, S., & Goel, R. K. (2019). Optimal pollution control in a mixed oligopoly with research spillovers. *Australian Economic Papers, 58*(1), 21–40.

Ioannidis, A., Chalvatzis, K., Leonidou, L., & Feng, Z. (2021). Applying the reduce, reuse, and recycle principle in the hospitality sector: Its antecedents and performance implications. *Business Strategy and the Environment, 30*(7), 3394–3410.

Jornal de Negócios. (2025). https://www.jornaldenegocios.pt/empresas/detalhe/governo-angolano-anuncia-privatizacao-de-hoteis-das-redes-iu-e-bina. (in Portuguese). Accessed 11 Apr 2025.

Leal, M., Garcia, A., & Lee, S.-H. (2018). The timing of environmental tax policy with a consumer-friendly firm. *Hitotsubashi Journal of Economics, 59*(1), 25–43.

Matsumura, T. (1998). Partial privatization in mixed duopoly. *Journal of Public Economics, 70*, 473–483.

Ouattara, K. S. (2019). Pollution abatement and partial privatization. *Economics Bulletin, 39*(3), 1887–1897.

Pal, R., & Saha, B. (2015). Pollution tax, partial privatization and environment. *Resource and Energy Economics, 40*, 19–35.

Wang, L. F. S., & Wang, J. (2009). Environmental taxes in a differentiated mixed duopoly. *Economic Systems, 33*(4), 389–396.

World Tourism Organization. (2015). *Affiliate members global reports, Volume eleven – Public-Private Partnerships: Tourism Development*. UNWTO. https://catedratim.wordpress.com/wp-content/uploads/2017/01/omt-2015-global_report_public_private_partnerships_tourism_development.pdf

Xing, M. Q., Tan, T. T., Xia, W., & Liu, L. L. (2020). Environmental R&D and privatization in a mixed duopoly market. *Managerial and Decision Economics, 41*(1), 93–105.

Xing, M., Yang, L., Tan, T., Wang, X., Li, Q., & Wang, B. (2022). Environmental taxes and the effects of partial privatization on environmental R&D, environment and welfare. *Economic Research-Ekonomska Istraživanja, 35*(1), 5164–5183.

Xu, L., Cho, S., & Lee, S.-H. (2016). Emission tax and optimal privatization in Cournot-Bertrand comparison. *Economic Modelling, 55*, 73–82.

Chapter 8
CSR and Quantity Leadership in Restaurant Competition

Fernanda A. Ferreira, Daniela Meneses, Sandro Carvalho, and Rita Peres

Abstract This paper investigates the critical role of Corporate Social Responsibility (CSR) in the food industry. It analyzes CSR's impact on competition between a CSR-focused restaurant and a traditional for-profit (FP) restaurant. The research employs a game theory model to examine how each restaurant chooses the quality and quantity of its products, influenced by its differing objectives (CSR vs. profit maximization). The study's main contributions include analyzing scenarios where a CSR restaurant competes with a traditional for-profit establishment, considering two leadership structures (CSR restaurant leading or FP restaurant leading the market), and determining equilibrium quality levels, quantities, profits, and overall social welfare in each model.

This paper investigates the effects of CSR within the food industry. We develop a model where a restaurant with a CSR focus competes with a traditional for-profit (FP) establishment in determining the quality and quantity of their offerings, in the order mentioned. Yet, while the levels of quality are chosen at the same time, the quantities are determined one after the other. Our analysis considers two leadership

F. A. Ferreira (✉)
UNIAG, CiTUR, ESHT, Polytechnic of Porto, Vila do Conde, Portugal
e-mail: faf@esht.ipp.pt

D. Meneses
UNIAG, ESHT, Polytechnic of Porto, Vila do Conde, Portugal
e-mail: dgm@esht.ipp.pt

S. Carvalho
ESHT, Polytechnic of Porto, Vila do Conde, Portugal
e-mail: ssc@esht.ipp.pt

R. Peres
CiTUR, Centre for Tourism Research, Development, and Innovation, Estoril Higher Institute for Tourism & Hotel Studies, Estoril, Portugal
e-mail: rita.peres@eshte.pt

A. L. Negrusa, M. M. Coroş (eds.), *Innovation, Sustainability, and Growth in a VUCA Environment*, Springer Proceedings in Business and Economics,
https://doi.org/10.1007/978-3-032-11639-0_8

structures: (i) the CSR restaurant leading the market; and (ii) the FP restaurant leading the market.

Keywords Food industry · Corporate social responsibility · Sustainable tourism · Leadership · Social welfare

8.1 Introduction

Recent reports from the World Health Organization highlight a concerning reality: the second Sustainable Development Goal—the eradication of hunger—remains far from being achieved. Global malnutrition rates have stagnated over the past 3 years, a trend further aggravated by the socio-economic repercussions of the COVID-19 pandemic. By 2023, it is estimated that between 713 and 757 million people worldwide may be affected by hunger, equating to approximately one in every 11 individuals globally. This figure is even more alarming in Africa, where the prevalence rises to one in five (FAO et al., 2024). In this context, the food industry has a crucial role to play in tackling these challenges through the adoption of innovative and ethically responsible practices, the promotion of sustainable production methods, and a strengthened commitment to corporate social responsibility (CSR).

Although immediate hunger can be satisfied with low-quality food, the pursuit of food security must go beyond the mere availability of calories, focusing instead on access to nutritious and high-quality food to ensure long-term health and well-being. Food quality is not a simple concept; it is complex and multifaceted, varying across different situations and contexts. It can be assessed objectively, based on measurable characteristics such as chemical composition, but also subjectively, shaped by consumer expectations, perceptions, and preferences.

Food safety ensures that food is produced safely and free from harmful contaminants, thereby protecting public health, maintaining consumer confidence, supporting economic sustainability, and ultimately contributing to individual and community well-being (Nie & Chen, 2014; Chen et al., 2017). Barbosa and Pumpín (2024) found that companies investing in water management enjoy a stronger reputation among peers within the same sector, thereby enhancing their legitimacy and standing. According to Toukabri and Chaouachi (2025), from the perspective of the food industry consumer, economic and social sustainability exert a greater impact and influence, making it essential to consolidate and justify the values associated with premium cuisine.

The growth of sustainable tourism presents a significant opportunity for the food industry. As tourists increasingly seek eco-friendly dining experiences—including farm-to-table options and establishments that prioritize local, organic, and sustainable sourcing—there is a rising demand that encourages the food industry to adopt more sustainable practices. This shift directly contributes to the delivery of higher-quality dishes for tourists. Food quality is a multifaceted concept that encompasses all the characteristics and attributes of a food product which influence its value to the consumer. It is not solely about taste; rather, it is a holistic assessment that includes both objective and subjective dimensions. Ultimately, high food quality aims to meet or exceed consumer expectations, offering a positive and satisfactory experience while ensuring safety and nutritional value.

Researchers have increasingly focused on quality competition in the food industry. For example, Chen et al. (2018) employed a two-stage dynamic duopoly model to analyse how reputation and quality influence competition between restaurants, with quality decisions preceding quantity choices (see also Ferreira et al., 2022). Building on this type of framework, Ferreira et al. (2021) examined a similar competitive environment, but with restaurants choosing prices rather than quantities after setting quality levels.

Nie and Chen (2014) examined how increasing competition (i.e., the number of firms) influences output, prices, and profits in a food industry oligopoly, where identical products compete in terms of quantity. Their findings indicated that greater competition reduced individual firm output and prices, diminished profits per firm, but enhanced social welfare. Yang and Nie (2016) explored how product substitutability affects asymmetric competition in the restaurant industry. They employed a two-stage game model to determine equilibrium outcomes under both Cournot and Stackelberg competition. Chen and Nie (2016) analysed the impact of Corporate Social Responsibility (CSR) in a mixed duopoly within the restaurant sector, in which one firm was profit-oriented while the other integrated CSR principles. Their model considered simultaneous decisions on quality, followed by quantity choices. The results showed that CSR adoption led to higher quality and quantity levels for the CSR-oriented firm, but the opposite for the profit-maximizing competitor.

CSR has been gaining significant importance in gastronomic tourism. In fact, the relevance of CSR in the food industry can be observed across various factors:

- *Builds Consumer Trust*: Consumers are increasingly concerned about the origins of their food, how it is produced, and the ethical practices of food companies. CSR initiatives—such as sourcing sustainable ingredients, ensuring fair labor practices, and maintaining transparency in production processes—can greatly enhance consumer trust.
- *Promotes Sustainability*: The food industry has a considerable environmental footprint. CSR promotes practices that reduce negative environmental impacts, including waste reduction, water and energy conservation, and support for sustainable agriculture.
- *Upholds Ethical Practices*: CSR involves maintaining high ethical standards in all areas of business, including fair treatment of workers, responsible sourcing, and honest marketing.
- *Encourages Innovation*: The pursuit of CSR objectives can foster innovation within the food industry. This may include developing new sustainable packaging, creating healthier food options, or adopting more efficient and ethical production methods.

The complex global challenge of climate change, characterized by widespread weather alterations and rising global temperatures, has emerged over recent decades. This phenomenon is expected to have a significant impact on both the availability and safety of our food supply (Duchenne-Moutien & Neetoo, 2021). Acknowledging the seriousness of this issue, the FAO has published key reports, including the foundational Climate Change: Implications for Food Safety in 2008 (FAO, 2008), and the more recent Climate Change: Unpacking the Burden on Food Safety in 2020 (FAO, 2020), which builds on the growing scientific understanding of the links between climate change and foodborne hazards.

The role of CSR within the tourism sector has been the subject of inquiry for numerous scholars (Font & Lynes, 2018). In exploring the impact of CSR on customer relationships within the food industry, Vasilev's (2025) study provides valuable insights. The research investigates how CSR strategies and practices foster consumer trust, satisfaction, and loyalty, acknowledging the growing consumer demand for ethical and sustainable business operations. It identifies key consumer expectations regarding social and environmental responsibility and analyses effective CSR initiatives implemented by leading companies in the sector. The study further evaluates the impact of these initiatives on customer engagement and the development of long-term relationships.

Biggi et al. (2024) examined the connection between CSR initiatives and state regulatory interventions designed to reduce food loss and waste. Their research revealed that food companies actively pursuing a CSR strategy, especially those that showcase their environmental and social achievements in their CSR reports, are more likely to successfully reduce food loss and waste.

Building on the work of Kong (2012) and Chen and Nie (2016) regarding CSR in the food industry, this study adopts their theoretical model to analyze the effects of CSR on key outcomes such as food quality, quantity, firms' profits, and consumer surplus. The model developed by Chen and Nie (2016) assumes that both firms initially determine product quality levels and subsequently select quantities. Ferreira et al. (2024), however, approached the market differently, treating quantities as the primary decision variable. This paper explores similar issues, but considers a sequential decision-making process for quality levels, rather than a simultaneous one.

Food industry managers can find essential guidance in Eledum and Elmahgop's (2024) research. It shows them how to integrate CSR into their core business strategies, foster innovation, and strengthen stakeholder relationships to achieve sustainable growth and profitability. The researchers' findings reveal that CSR significantly improves financial performance and competitive advantage, especially when companies focus on innovation.

Our study gives us some results that we can highlight, such as the Impact on CSR restaurant, impact on FP restaurant, overall industry impact, benefits for consumers and society, consumer surplus, and social welfare, which will be detailed below.

Impact on CSR restaurant: The study concludes that the presence of CSR positively influences the CSR restaurant. It leads to an enhancement of both the quality and the quantity of food offered by the CSR-focused establishment. This could be due to the CSR restaurant's commitment to providing better, more sustainable options and potentially attracting a larger customer base due to its values.

Impact on for-profit restaurant: Conversely, the study finds that the competition from a CSR restaurant leads to a diminishment in both the quality and the quantity of food offered by the for-profit restaurant. This could be a strategic response to the CSR competitor, perhaps focusing on cost-cutting or targeting a different segment of the market.

Overall industry impact: Importantly, the study suggests that CSR has a positive impact on the food industry. It leads to an enhancement of both the overall quality and the overall quantity of food available in the market. This implies that the positive effects of the CSR restaurant outweigh the negative adjustments made by the FP restaurant.

Benefits for consumers and society: The research further concludes that CSR enhances both consumer surplus and social welfare.

Consumer surplus: This refers to the benefit consumers receive from purchasing goods or services at a price lower than their maximum willingness to pay. Higher quality and potentially greater availability of food due to CSR likely contribute to increased consumer surplus.

Social welfare: This is a broader measure of societal well-being, considering factors like consumer benefits, producer profits, and environmental impacts. The positive effects of CSR on quality, quantity, and potentially sustainability contribute to higher social welfare.

The paper distinguishes itself from previous research in its methodological approach, particularly in the sequencing of decisions within the game theory model.

Sequential vs. *simultaneous quality decisions*: While Chen and Nie (2016) examined a model where both firms simultaneously determined product quality levels and then quantities, this study introduces novelty by considering that restaurants choose quality levels sequentially, rather than simultaneously.

Quantity as primary decision variable: Unlike Ferreira et al. (2024), who modeled the market with quantities as the primary decision variable, and Ferreira et al. (2021) and Chen et al. (2018), who had quality decisions preceding quantity choices, this paper's model structure, where quality levels are chosen simultaneously but quantities are determined one after the other, with different leadership scenarios, is a key differentiator.

Leadership structures in quantity decisions: The study specifically analyzes two distinct scenarios based on which restaurant acts as the "leader" in the quantity decision-making process: (i) the CSR restaurant leading, and (ii) the FP restaurant leading, after an initial simultaneous quality setting. This detailed examination of sequential quantity decisions under different leadership appears to be a novel aspect.

Building on existing frameworks with a twist: The paper builds on market structures outlined by Chen and Nie (2016) and work by Kong (2012) and Chen and Nie

(2016), but diverges by assuming a sequential, rather than simultaneous, decision-making process for quantity.

This paper proceeds as follows: The model is established in the subsequent section. Following this, we describe the model, present our analysis and findings, and offer a comparison of sequential versus simultaneous decisions. The final section provides the conclusion.

8.2 Research Purpose and the Methodological Approach

The core purpose of this paper is to highlight the crucial role of CSR in the food industry and to analyze its impact on competition between a CSR-focused restaurant and a traditional FP restaurant. It also aims to explore the effects of CSR within the food industry by examining how two different types of restaurants make decisions about the quality and quantity of their offerings.

The objectives of the research are to analyze a scenario where a restaurant committed to CSR competes with a traditional for-profit restaurant; to model how each restaurant chooses the quality and quantity of its products, considering its differing objectives (CSR vs. profit maximization); to investigate the effects of CSR within the food industry; to develop a model where a CSR-focused restaurant competes with a traditional for-profit establishment in determining the quality and quantity of their offerings; to consider two leadership structures: (i) the CSR restaurant leading the market; and (ii) the FP restaurant leading the market; to determine the equilibrium quality levels and quantities, as well as the resulting profits for the restaurants and overall social welfare, in each model; and to analyze how the level of CSR influences the various outcomes of the model.

We analyze a scenario where a restaurant with a strong commitment to CSR competes directly with a traditional for-profit restaurant. The model focuses on how each restaurant chooses the quality and quantity of its products. These are likely influenced by their differing objectives (CSR versus profit maximization). We utilize game theory, a mathematical framework for analyzing strategic interactions between rational decision-makers. The game theory model provides a framework for understanding the strategic interactions between different types of food businesses in the context of CSR. This allows us to model how each restaurant's choices are influenced by the anticipated actions of the other. The paper examines two distinct scenarios based on which restaurant acts as the "leader" in the market:

(i) CSR restaurant as leader: After a simultaneous decision on quality levels, this implies the CSR restaurant makes its quantity decisions first, and the FP restaurant responds accordingly.
(ii) FP restaurant as leader: In this case, also after a simultaneous decision on quality levels, the FP restaurant sets its quantity first, and the CSR restaurant adapts its strategy.

Building on the market structure outlined by Chen and Nie (2016), our model diverges by assuming a sequential, rather than simultaneous, decision-making process for quantity. In the model studied by Ferreira, decisions about quality levels were made sequentially, whereas the quantities were chosen simultaneously.

We examine market competition between two distinct restaurants: R_1, characterized by its integration of CSR into its objectives, and R_2, driven by profit maximization. Throughout this paper, we will refer to R_1 as the CSR restaurant and R_2 as the for-profit (FP) restaurant. The representative consumer maximizes the utility function

$$U(q_1, q_2, x_1, x_2) - p_1 q_1 - p_2 q_2,$$

where q_i is the quantity supplied by restaurant R_i, x_i is the quality level of restaurant R_i's offering, and p_i is the price set by restaurant R_i, for $i = 1, 2$. The linear inverse demand function for each restaurant R_i is given by:

$$p_i = 2x_i - q_i - \gamma q_j,$$

where γ is a measure of how substitutable the products offered by the two restaurants are, with $i, j = 1, 2, i \neq j$. Consequently, the direct demand function for restaurant R_i is expressed as:

$$q_i = \frac{1}{1-\gamma^2}\left(\alpha x_i - \alpha\gamma x_j - p_i + \gamma p_j\right).$$

The profit of restaurant R_i, denoted as π_i, is calculated as:

$$\pi_i = (p_i - x_i) q_i - \tau\left(x_i^2 - x_i\right).$$

In this formula, the cost of producing a food quantity q_i at quality level x_i is $q_i x_i$, and the fixed costs associated with quality level x_i are represented by $\tau\left(x_i^2 - x_i\right)$.

Assumption 1: To facilitate a simpler analysis, we adopt the parameter values $\gamma = \tau = 0.5$.

The CSR-focused restaurant R_1 aims to maximize $V = \pi_i + \theta CS$, reflecting its integration of CSR into business operations. This objective function includes its profit (π_1) and a weighted consumer surplus (θCS), where the CSR degree θ falls within the range [0, 1]. The consumer surplus (CS) is defined as $CS = \frac{1}{2}\left(q_1^2 + q_2^2\right) + \gamma q_1 q_2$. In contrast, the for-profit (FP) restaurant R_2 seeks to maximize its standard profit function, π_2.

The measure of social welfare, W, is defined as the addition of the profit earned by restaurant R_1, the profit earned by restaurant R_2, and the resulting consumer surplus:

$$W = \pi_1 + \pi_2 + CS.$$

The methodology employs a three-stage game theory model to analyze non-cooperative competition. In both cases, the first stage involves simultaneous quality setting by both restaurants. The cases diverge in the subsequent quantity decisions:

- *Case I*: After simultaneously setting quality, the CSR restaurant chooses its quantity, followed by the FP restaurant's quantity choice;
- *Case II*: After simultaneously setting quality, the FP restaurant chooses its quantity, followed by the CSR restaurant's quantity choice.

In each model, we explicitly determine the equilibrium quality levels and quantities. Furthermore, we calculate the resulting profits for the restaurants and the overall social welfare. We also analyze how the level of CSR influences the various outcomes of the model.

8.3 Results I: CSR Restaurant as Leader

Let's break down this marketing competition using a three-stage game:

- First, both restaurants simultaneously determine the quality levels of their products, x_1 and x_2;
- Next, the CSR restaurant determines the quantity it will offer, q_1;
- Finally, the FP restaurant observes CSR restaurant's choice and then sets its own quantity, q_2.

The model presented above is analyzed and solved via backward induction.

Analyzing the final stage's optimality condition, $\partial\pi_2/\partial q_2 = 0$, results in the following solution:

$$q_2 = \frac{2x_2 - q_1}{4}. \tag{8.1}$$

Substituting (Eq. 8.1) into the objective function V for restaurant R_1, and maximizing with respect to q_1 ($\partial V/\partial q_1 = 0$), we obtain the following expression for q_1:

$$q_1 = \frac{2\left(8x_1 - (2-\theta)x_2\right)}{28 - 13\theta}. \tag{8.2}$$

Then,

$$q_2 = \frac{-4x_1 + (15 - 7\theta)x_2}{28 - 13\theta}. \tag{8.3}$$

In the initial stage, CSR restaurant R_1 determines its optimal quality level x_1 by maximizing its objective function V. At the same time, FP restaurant R_2 decides on its ideal quality level x_2 to make the most profit π_2. To find this equilibrium, we substitute Eqs. (8.2) and (8.3) into the definition of V and π_2. Then, by solving the first-order conditions, $\partial V/\partial x_1 = 0$ and $\partial \pi_2/\partial x_2 = 0$, we arrive at the equilibrium quality levels for restaurants R_1 and R_2:[1]

$$x_1^I = \frac{222 - 200\theta + 45\theta^2}{2\left(126 - 212\theta + 71\theta^2\right)}, x_2^I = \frac{216 - 464\theta + 169\theta^2}{2\left(126 - 212\theta + 71\theta^2\right)}. \tag{8.4}$$

By inserting (Eq. 8.4) into Eqs. (8.2) and (8.3), the equilibrium quantities become:

$$q_1^I = \frac{48 + 6\theta - 13\theta^2}{126 - 212\theta + 71\theta^2}, q_2^I = \frac{84 - 235\theta + 91\theta^2}{2\left(126 - 212\theta + 71\theta^2\right)}.$$

Assumption 2: To ensure interior solutions exist, we assume $\theta < 3/7$.

Proposition 1. The results demonstrate that the CSR restaurant excels in both food quality and quantity when compared with the FP restaurant.

The profits for restaurant R_1 (π_1) and restaurant R_2 (π_2) are given by:

$$\pi_1^I = \frac{22788 - 71856\theta + 65944\theta^2 - 23954\theta^3 + 3039\theta^4}{8\left(126 - 212\theta + 71\theta^2\right)^2},$$

$$\pi_2^I = \frac{21888 - 87024\theta + 122718\theta^2 - 66252\theta^3 + 11999\theta^4}{8\left(126 - 212\theta + 71\theta^2\right)^2}.$$

Moreover, the consumer surplus (CS) and social welfare (W) are expressed as:

$$CS^I = \frac{24336 - 58728\theta + 69397\theta^2 - 36192\theta^3 + 6591\theta^4}{8\left(126 - 212\theta + 71\theta^2\right)^2},$$

$$W^I = \frac{69012 - 217608\theta + 258059\theta^2 - 126398\theta^3 + 21629\theta^4}{8\left(126 - 212\theta + 71\theta^2\right)^2}.$$

A comparative static analysis will now be conducted to evaluate the impact of CSR adoption on market equilibrium outputs. Through straightforward calculations, we arrive at the subsequent findings.

---

[1] Throughout the paper we use the notation superscript I to refer to the CSR restaurant leader case.

Proposition 2. Greater CSR involvement boosts both the quality and quantity of food at the CSR restaurant but reduces these at the FP restaurant.

So, we can say that when a CSR-focused restaurant increases its commitment to CSR (perhaps by investing more in sustainable practices, ethical sourcing, or community support), it benefits in two ways: (a) Food quality improves, likely because of better sourcing, attention to health, or customer satisfaction; (b) Food quantity increases, perhaps because CSR attracts more customers (increased demand), or more resources become available to serve more food. However, the same increase in CSR involvement has a negative effect on traditional for-profit restaurants: (a) Quality drops, possibly because the FP restaurant cannot or will not match the CSR efforts (e.g., they may cut costs to compete); (b) Quantity also drops, possibly because they lose custom to CSR restaurants or have fewer resources to produce the same amount of food. So, increased CSR creates competitive pressure, and if FP restaurants do not adapt, they suffer.

Proposition 3. The analysis also reveals that as CSR engagement grows, the differences in both food quality and quantity become more pronounced, and the total quantity of food also tends to increase.

According to this proposition, higher CSR levels drive an increase in the quantity of offerings, whilst simultaneously making the differences between companies with varying CSR commitments more apparent.

Proposition 4. While the FP restaurant's profits tend to decrease as the CSR restaurant invests more in CSR, there is an inverse U-shaped relationship between the CSR restaurant's own level of CSR engagement and its profitability.

The findings indicate an ambiguous effect of increased CSR engagement on a CSR restaurant's profitability. However, greater social responsibility in CSR restaurants, conversely, leads to profit losses for traditional FP restaurants.

Let us now expand the interpretation and commentary of empirical results in this scenario, in which, after both restaurants simultaneously set their quality levels, the CSR restaurant makes its quantity decision first, followed by the FP restaurant.

The study concludes that the CSR restaurant, by leading with its CSR commitment, experiences a positive influence on its operations. This leads to an improvement in both the quality and quantity of food offered by the CSR-focused establishment. This outcome is likely attributed to the CSR restaurant's dedication to providing superior, more sustainable options, which in turn can attract a larger customer base due to its alignment with consumer values.

Conversely, when a CSR restaurant is present and takes a leading role, it results in a reduction in both the quality and quantity of food provided by traditional for-profit restaurants. This could be a strategic response from the FP competitor, possibly involving cost-cutting measures or a shift in focus to a different market segment to compete with the CSR-driven establishment.

Importantly, the study suggests that CSR has a net positive impact on the food industry. The positive effects observed in the CSR restaurant, in terms of increased quality and quantity, appear to outweigh the negative adjustments made by the FP restaurant. This implies that CSR drives an overall improvement in the quality and quantity of food available in the market.

The research further concludes that CSR significantly enhances both consumer surplus and social welfare. This indicates that consumers benefit more, as they are able to purchase goods or services at a price lower than their maximum willingness to pay. The higher quality and potentially greater availability of food, driven by CSR, is likely to contribute to this increased consumer surplus. The positive effects of CSR on food quality, quantity, and potential sustainability efforts contribute to a higher overall social welfare.

8.4 Results II: FP Restaurant as Leader

We will now explore a marketing competition by applying the following three-stage game model:

- First, both restaurants simultaneously choose their product quality levels;
- Knowing the quality levels, the FP restaurant then decides on its offering quantity;
- Observing the FP restaurant's quantity, the CSR restaurant sets its own quantity.

Applying the same analytical approach as in the previous section yields the following equilibrium outcomes:[2]

$$x_1^{II}=\frac{3\left(6-5\theta+\theta^2\right)}{21-35\theta+12\theta^2},x_2^{II}=\frac{111-278\theta+183\theta^2-36\theta^3}{2\left(63-126\theta+71\theta^2-12\theta^3\right)},$$

$$q_1^{II}=\frac{21+4\theta-19\theta^2+6\theta^3}{(3-\theta)\left(21-35\theta+12\theta^2\right)},q_2^{II}=\frac{4\left(6-19\theta+14\theta^2-3\theta^3\right)}{(3-\theta)\left(21-35\theta+12\theta^2\right)},$$

$$\pi_1^{II}=\frac{1368-5019\theta+6323\theta^2-3638\theta^3+942\theta^4-63\theta^5-9\theta^6}{2(3-\theta)^2\left(21-35\theta+12\theta^2\right)^2},$$

$$\pi_2^{II}=\frac{1899-8211\theta+13441\theta^2-9945\theta^3+3384\theta^4-432\theta^5}{8(3-\theta)\left(21-35\theta+12\theta^2\right)^2},$$

[2] Throughout the paper we use the notation superscript *II* to refer to the FP restaurant leader case.

$$CS^{II} = \frac{1521 - 49800\theta + 80980\theta^2 - 74280\theta^3 + 38010\theta^4 - 10080\theta^5 - 1080\theta^6}{2(3-\theta)^2(21-35\theta+12\theta^2)^2},$$

$$W^{II} = \frac{3(1917 - 6753\theta + 9551\theta^2 - 6543\theta^3 + 2160\theta^4 - 276\theta^5)}{8(3-\theta)(21-35\theta+12\theta^2)^2}.$$

Unlike the previous scenario, if the FP restaurant takes the lead, it can achieve higher production and quality levels, particularly when the other restaurant's commitment to social responsibility is limited. Similarly, the remaining qualitative outcomes align with those observed when the CSR restaurant held the leading position.

Let us now expand the interpretation and commentary of empirical results in this scenario in which, after both restaurants simultaneously set their quality levels, the FP restaurant makes its quantity decision first, followed by the CSR restaurant.

Unlike the previous scenario, if the FP restaurant takes the lead, it can achieve higher production and quality levels. This advantage is particularly noticeable when the CSR restaurant demonstrates a limited commitment to social responsibility. This suggests that a strong profit-driven leadership can dominate the market if the CSR competitor is not as heavily invested in its social mission.

Despite the change in leadership, the study found that the remaining qualitative outcomes largely align with those observed when the CSR restaurant held the leading position. This implies that while leadership dictates who gains the initial advantage in terms of production and quality, the underlying competitive dynamics and their broader implications for the industry and society remain similar in nature.

8.5 Comparisons

Here, we present a comparison of selected results derived from the two models. Based on the quantitative results, we can establish the following:

- The aggregate quantity in the market is higher when the leader is the CSR restaurant.
- A restaurant achieves better quality in its production by taking on a leadership role.
- A leader restaurant's commitment to social responsibility enhances social welfare.

The comparison of results from both models reveals important insights. The total quantity of food in the market is higher when the CSR restaurant assumes the leadership role. This reinforces the idea that CSR, when championed by a market leader, can lead to a greater overall supply of food.

Furthermore, a restaurant achieves better quality in its production when it takes on a leadership role, regardless of whether it's the CSR or FP restaurant. This suggests that the strategic advantage of being a first-mover in quantity decisions allows the leader to optimize its quality more effectively.

Crucially, the study finds that a leader restaurant's commitment to social responsibility enhances overall social welfare. This highlights the significant societal benefit derived when a market leader, especially one with CSR principles, influences market dynamics.

8.6 Conclusions

Employing a dynamic game theory model, this study investigated CSR strategies in a duopolistic food industry. We considered two distinct scenarios: the first where a CSR-focused restaurant acted as the market leader, and the second where a for-profit restaurant held the leadership position.

Results from the first scenario highlight the CSR restaurant's advantage in food quality and quantity over the FP restaurant. Conversely, should the FP restaurant assume a leading position, it can attain elevated levels of production and quality, particularly in circumstances where the other establishment demonstrates a limited commitment to social responsibility. We also established that when a restaurant takes the lead, it elevates the quality of its production. Furthermore, the commitment to social responsibility demonstrated by a leading restaurant contributes to the enhancement of social welfare.

The paper's exploration of different leadership scenarios (CSR vs. FP restaurant leading) highlights that market dynamics vary. Internationally, this means that policy approaches to fostering CSR in the food sector might need to be adapted to the specific competitive landscapes of different countries or regions. In markets dominated by large multinational corporations, different strategies might be required compared to those with a prevalence of small, local businesses.

In conclusion, this research suggests that the integration of CSR principles in the food industry, even in a competitive environment, can lead to positive outcomes for both consumers and society by encouraging higher quality and quantity of food, particularly from CSR-focused businesses, and ultimately pushing the entire industry towards more responsible practices.

Furthermore, the empirical findings of this study provide valuable policy insights for governments and regulatory authorities seeking to encourage sustainable and responsible practices in the food sector. These policies may (i) involve promoting the adoption of CSR in the food industry through incentives for CSR initiatives and public recognition programs; and (ii) leverage CSR to enhance social welfare by integrating it into food security strategies and consumer awareness initiatives.

Acknowledgements This work was supported by national funds through FCT/MCTES (PIDDAC): UIDB/04752—Unidade de Investigação Aplicada em Gestão (UNIAG).

References

Barbosa, M. W., & Pumpín, M. d. l. Á. R. (2024). The effects of water footprint management on companies' reputations and legitimacy under the influence of corporate social responsibility and government support: Contributions to the Chilean agri-food industry. *Water, 16*(19), 2–19.

Biggi, G., Principato, L., & Castellacci, F. (2024). Food waste reduction, corporate responsibility and national policies: Evidence from Europe. *British Food Journal, 126*(13), 470–485.

Chen, Y.-H., & Nie, P.-Y. (2016). Duopoly competition in food industry considering corporate social. *Quality Assurance and Safety of Crops & Food, 8*(1), 33–40.

Chen, Y.-H., Nie, P.-Y., & Yang, Y. C. (2017). Effects of corporate social responsibility on food safety. *Agricultural Economics – Czech, 63*, 539–547.

Chen, Y.-H., He, Q., & Paudel, K. P. (2018). Quality competition and reputation of restaurants: The effects of capacity constraints. *Economic Research – Ekonomska Istrazivanja, 31*(1), 102–118.

Duchenne-Moutien, R. A., & Neetoo, H. (2021). Climate change and emerging food safety issues: A review. *Journal of Food Protection, 84*(11), 1884–1897.

Eledum, H., & Elmahgop, F. O. (2024). The effect of corporate social responsibility on corporate performance in the food industry in Saudi Arabia: A partial least squares structural equation modeling approach. *Sustainability, 16*, 7925.

FAO. (2008). *Climate change: Implications for food safety*. FAO. http://www.fao.org/3/i0195e/i0195e00.pdf. Accessed 27 Aug 2024

FAO. (2020). *Climate change: Unpacking the burden on food safety* (Food Safety and Quality Series No. 8). FAO. https://www.fao.org/3/ca8185en/CA8185EN.pdf. Accessed 03 Mar 2025.

FAO, IFAD, UNICEF, WFP and WHO. (2024). *The state of food security and nutrition in the world 2024 – Financing to end hunger, food insecurity and malnutrition in all its forms*. FAO.

Ferreira, F., Ferreira, F., & Fernandes, P. O. (2021). Price and quality competition in the restaurant industry: Effects of restaurants' reputation. *Revista Turismo & Desenvolvimento, 37*, 21–29.

Ferreira, F. A., Ferreira, F., & Fernandes, P. O. (2022). Sequential quality competition in the restaurant industry: Effects of restaurants' reputation. Springer proceedings in business and economics. In V. Katsoni & A. C. Şerban (Eds.), *Transcending Borders in tourism through innovation and cultural heritage* (pp. 779–789). Springer.

Ferreira, F. A., Ferreira, F., & Santos, A. M. (2024). Corporate social responsibility and quality leadership in food industry. In J. V. Carvalho et al. (Eds.), *Advances in tourism, technology and systems, ICOTTS'24, Madeira, Portugal*. Springer.

Font, X., & Lynes, J. (2018). Corporate social responsibility in tourism and hospitality. *Journal of Sustainable Tourism, 26*(7), 1027–1042.

Kong, D. M. (2012). Does corporate social responsibility matter in the food industry? Evidence from a nature experiment in China. *Food Policy, 37*, 323–334.

Nie, P.-Y., & Chen, Y.-H. (2014). A general equilibrium analysis of food industry considering the food quality. *Agricultural Economics-Zemedelska Ekonomika, 60*, 301–308.

Toukabri, M., & Chaouachi, M. (2025). Exploring the influence of corporate social responsibility, Blockchain transparency, and cultural alignment on consumer trust and premium pricing willingness for local food in Saudi Arabia. *Current Research in Nutrition and Food Science, 13*(1), 304–316.

Vasilev, V. (2025). Corporate social responsibility in the food industry: Challenges and prospects for improving customer relationships. *BIO Web of Conferences, 173*, 04001.

Yang, Y.-C., & Nie, P.-Y. (2016). Asymmetric competition in food industry with product substitutability. *Agricultural Economics – Czech, 62*(7), 324–333.

Chapter 9
Influence of Comfort Level and Location on Accommodation Units

Violeta Strătila Chele and Aurel Costel Chele

Abstract The number of stars is an indicator of the quality of the services and facilities available in an accommodation unit. The standardization of quality in tourism, by introducing the number of stars, professionalizes tourism and leads to a more competitive and better-regulated industry.

The facilities and standards of the services offered and the level of comfort available to the accommodation unit are the defining elements in the construction of the star rating of the hotels. The correlation between the standard that a hotel assumes and the number of stars, a correlation previously verified by the authorities with competencies in the field of tourism, gives confidence to the contemporary tourist.

The chapter aims to investigate the influence exerted by the level of comfort and geographical location on the dynamics of the number of accommodation units, as well as on their economic performance. The research methods used are the statistical analysis of data from sites such as the National Institute of Statistics, the Ministry of Tourism, and the Ministry of Public Finance, as well as the comparative study between types of accommodation units and location.

The indicators used in the analysis include both variables specific to the tourism sector—such as the number of accommodation units, the volume of arrivals in tourist accommodation facilities, and the occupancy rate—and economic performance indicators, such as turnover, level of expenses, profit obtained, and rate of return. For the analysis of the economic performance indicators, a sample of accommodation units from the spa resort of Slănic Moldova and the municipality of Bacău was constituted, the selection being made based on comparable criteria, such as the type of accommodation unit, the classification category, and the accommodation capacity.

The results of the research lead to the existence of a positive impact generated by the tourist classification on the economic performance of accommodation units, emphasizing that this influence is complemented by other determining factors in

V. S. Chele (✉) · A. C. Chele
Ştefan cel Mare University of Suceava, Suceava, Romania

A. L. Negrusa, M. M. Coroş (eds.), *Innovation, Sustainability, and Growth in a VUCA Environment*, Springer Proceedings in Business and Economics,
https://doi.org/10.1007/978-3-032-11639-0_9

achieving important levels of performance indicators. As regards the influence of geographical location on the indicators of accommodation units, the research conducted did not identify a significant impact.

Keywords Tourist classification · Number of stars · Performance indicators · Accommodation units · Tourist destinations

9.1 Introduction

From a tourism perspective, Romania can be declared a blessed country, with assets from divine landscapes to medieval castles, from memorable churches to beautiful traditions. In this global market, the existence of tourism has consecrated, over the years, certain strong poles and certain areas dedicated to tourism, due to some given or constructed attributes.

Contemporary Romanian tourism is a very dynamic one and one that presents many overall perspectives; it is one of the pillars that support economic development. Therefore, the investigation of the tourism phenomenon, from multiple perspectives, becomes an essential necessity for the understanding and sustainable development of this sector.

Over the millennia, tourism has experienced a constant evolution, adapting to the needs and transformations of human society. From religious pilgrimages to spa practices, from educational to mass tourism, tourism has travelled a trajectory marked by diversification and standardization. Today, supported by technology and digitalization, tourism reflects contemporary values and culture, becoming an expression of modernity. The modernization of tourism was possible through its standardization.

The beginning of the twentieth century brought a new but particularly crucial element for improving the quality of tourist services. For the first time, the star system is introduced in the United Kingdom, its role being one that can standardize the services and quality of accommodation units for tourists. This innovation took international tourism to another major step, and later, other countries adopted this system.

The essence of creating a classification system for tourist establishments played a decisive role in their correct evaluation; of course, each country develops its own classification systems. Thus, tourism has become more organized; this star classification accurately describes the quality of hotels and tourist services. The organization of tourism by star classification propelled this sector, especially after World War II, when the middle layer of societies began to travel, thus generating mass tourism.

Globalization has taken over this five-star rating system, the specification being that certain criteria may vary from one country to another, so there are certain different modern criteria, such as the size of the rooms, their equipment, and the quality of the furniture. Globalization is the one that, through branched technology, has managed to drive tourism into other dimensions. The classification of tourism, the internet, the emergence of low-cost transport operators, the maturity of tour operators, and much more are factors that have led to what we see around us today, modern tourism. The above elements are the ones that gave people the incentive to travel; they are the ones that ensured a safety component of tourism, so that people became familiar with this concept, accepted it, and today's tourism is part of people's lives and becoming a culture of modern life.

Tourism classified by stars/flowers is one that ensures a standardized level of quality, this standardization having a special role in the development of competition. And it is the competition created within this economic sector that has generated new levels of quality, for the benefit of all tourists. The Romanian legislation on the classification of tourist accommodation units (Just.ro, 2025) provides for the priority role of this classification. It is specified that a classification of tourist reception structures has as a priority the protection of tourists, constituting a codified form of synthetic presentation of the level of comfort and the offer of services.

The Romanian literature on the correlation between tourism classification and the performance of a tourism business is limited.

9.2 Literature Review

Tourism is a sector of economic importance for most developed and developing countries. Costea et al. (2016) consider that, in this context, the tourism industry in different countries has started to make sustained efforts, in terms of both human resources and material, financial, and informational resources, to develop competitiveness in tourism activity at the international level.

Tourism is a sector of the economy in a continuous dynamic, but it is fragile in crisis situations. Hapenciuc et al. (2009) present aspects related to the fragility of the tourism market at the onset of a global crisis. They believe that the fragility of the tourism market makes it impossible to estimate the evolution of the hospitality industry for the coming years.

A country, even if it does not have strong tourism poles, must make the most of the tourism part. Boghean (2007) suggested that the tourism sector is one of the most important sources of wealth for nations, regardless of their level of development.

A certain number of stars of a tourist establishment has a certain level of quality, compared to another tourist organization that has a lower number of stars. At the same level of stars, competition makes its well-deserved place, the main goal of entrepreneurs in the tourist area being to hold a suitable position on the market. Verstiuc (2022) presents a hypothesis that the tendency to be the best on the market is always current. Thus, by creating offers, various strategies are used to promote the services offered by a hotel, but also to be one step ahead of its competitors with the same classification according to the number of stars of the hotel.

Contemporary tourism must have two main directions, namely, to be a quality one and a sustainable one. These two attributes of tourism are so well intertwined that modern tourism cannot function without one of them. In Ciobanu's (2022) opinion, quality is the defining and indispensable factor of sustainable tourism. To the question of how quality services can be provided in tourism, one of the answers comes from the same author—one of the ways to ensure the quality of the services provided, including in the field of tourism, is quality standards.

The standardization of Romanian tourism is applied according to Methodological norms of April 15, 2010, regarding the classification of tourist reception structures, so that the following types of tourist reception structures with accommodation functions can operate (Fig. 9.1):

The behaviour of the tourism entrepreneur, from the point of view of standardization, is directed towards quality, towards obtaining and preserving it. Even if fierce competition in tourism causes and generates a quality environment, this quality can still be affected by the behaviour of the decision-makers in the respective tourist destination. Suslenco and Movilă (2018) consider that measures such as the conservation and renovation of tourist attractions and the application of measures that would protect the environment are necessary to improve the activity.

In choosing a tourist destination, the natural landscape of the destination has a major influence. This aspect is also presented by Gyori (2011), finding the reason tourists spend their holidays in Chile: natural landscapes, unspoiled nature, exoticism, and culture. Our opinion about Romania is consistent with the above. Romania is a beautiful country with unique tourist destinations in the world.

Romania, through its history and special landscape, can and must better capitalize on national tourism. Abdul Basit et al. (2021) conclude that tourism and travel are an important cornerstone of countries' social and economic development. Also in the same paper, the authors express their opinion on the importance of the field of tourism, stating that tourism is the source of income, jobs, and recreation.

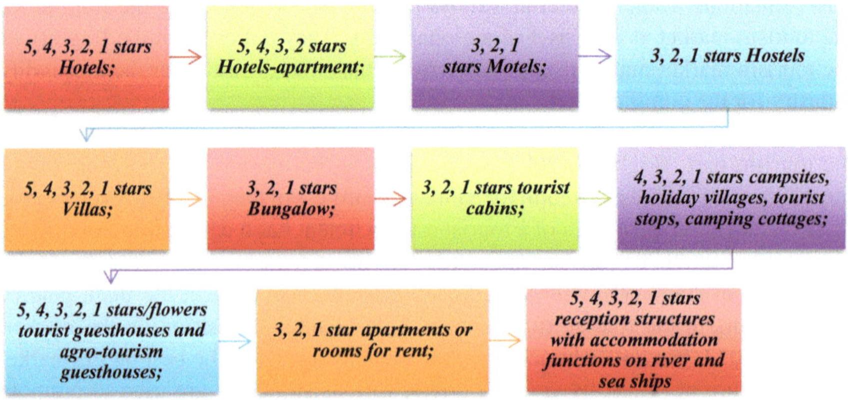

Fig. 9.1 Classification of accommodation structures. (Source: Adapted by the author according to (Just.ro, 2025))

As such, our regions can capitalize on natural beauty (mountains, lakes, forests, nature reserves, and beaches), but they can also capitalize on the exploration of the past (castles, museums, and monuments). The right combination of these can reposition Romania on the tourism map.

Of course, Romania's strategic development must not only impose the area of tourism. The diversification of all sectors of economic activities is a long-term economic solution for our regions.

Tourism can bring added value, but it also has a seasonal component, a fundamental problem, in terms of efficiency and business continuity. Ćorluka Goran et al. (2016) conclude that the problem of seasonal concentration of tourist demand is and will be a serious challenge for tour operators. The significant differences in occupancy of accommodation capacities between summer and winter cause inefficiency and unprofitable businesses.

Tourism businesses are affected by seasonal concentration, but more than that, efficiency must be properly checked even during seasonal periods. A survey of 84 hotels in Portugal's Algarve region (a region with an international reputation for high-quality golf courses) concludes that hotels without golf courses are more efficient. But Ricardo Oliveira de et al. are of the opinion that the Algarve's high reputation for golf is a factor in the tourism industry. Variables such as the number of stars of a hotel and those related to owning a golf course have a significant impact on hotels in the Algarve.

It is important to emphasize that obtaining a star rating for a hotel unit implies the existence of an adequate infrastructure. However, the efficiency of a tourism business must be evaluated not only in terms of the available infrastructure, correlated with the star rating, but also in relation to the general perception of the tourist destination at a macroeconomic level. Chele et al. (2024) argue that entrepreneurs in the tourism sector have capitalized on both the desire of local authorities to respond to the needs of the community and the tendency of people to seek socialization and diversion. Behind the cultural and artistic manifestations, there is also a significant economic interest.

Carles Mulet-Forteza et al. (2024) address, from a management perspective, the need for careful analysis of the available expansion strategies and their impact on the performance of hotel establishments, in both the short and long term. For decision-makers, the importance of choosing appropriate growth and diversification strategies is emphasized, as well as optimizing hotel characteristics—size, location, category, and range of services—to maximize operational efficiency and economic profitability.

9.3 Research Methodology

This research uses a quantitative approach, based on the comparative analysis of economic indicators specific to the tourism sector, to evaluate the performance of accommodation units according to the level of comfort and location.

For the empirical substantiation of the study, a sample of accommodation units located in the spa resort of Slănic Moldova and in the municipality of Bacău was selected, the choice of these locations being motivated by the differences in the tourist potential and the level of attractiveness, thus being able to identify a possible influence of the location in the evolution of the performance indicators of the accommodation units. The selection criteria of the analysed units included: the type of tourist reception structure, the classification category, the number of accommodation spaces, and the occupancy rate.

The analysed indicators are grouped into two categories:

- Indicators specific to the tourism sector—the number of accommodation units, the number of arrivals in tourist reception structures, and the occupancy rate
- Economic performance indicators—turnover, total costs, profit obtained, and degree of profitability

Data collection was conducted from official sources, such as the National Institute of Statistics, the Ministry of Tourism, and the Ministry of Public Finance, for the period 2021–2024. The information was processed and interpreted to identify relevant relationships between comfort level, location, and economic performance.

For the analysis of the performance indicators of companies with tourist activity, seven accommodation units from the two localities (seven units from Bacău municipality and seven units from Slănic Moldova) were selected, having similar selection criteria (type of unit, category, and number of accommodation spaces).

The chapter has certain research limitations, generated by the small size of the analysed sample, an aspect that can influence the degree of generalization of the results. Future research directions will consider expanding the sample of accommodation units to obtain a more comprehensive and relevant picture of the relationship between the level of comfort, location, and economic performance.

9.4 Results and Discussions

The accommodation units are divided into comfort categories, depending on the facilities and services offered, according to Table 9.1.

At the national level, **the number of accommodation units** is higher in the county capitals, with 2182 units in 2024, compared to 937 units in spa resorts. The higher number of accommodation units in the county capitals, compared to the spa resorts, is explained by the diversity of tourist demand, the developed infrastructure, the frequency of economic and cultural events, as well as the easier access to the qualified workforce.

In both tourist destinations, there are annual increases in the number of accommodation units classified with 4 or 5 stars/flowers, which shows an increased interest in the quality of the services offered.

Table 9.1 Accommodation units in Romania by categories and tourist destinations, 2021–2024

Comfort categories	Tourist destinations	Periods			
		Year 2021	Year 2022	Year 2023	Year 2024
Total	**Total**	***11,736***	***12,201***	***12,697***	***13,538***
	Spa resorts	*813*	*890*	*935*	*937*
	Bucharest and its county seats, except Tulcea	*1983*	*1981*	*2052*	*2182*
5 stars	**Total**	**144**	**152**	**204**	**175**
	Spa resorts	5	7	8	8
	Bucharest and its county seats, except Tulcea	26	25	33	38
4 stars	**Total**	**1194**	**1238**	**1265**	**1321**
	Spa resorts	63	67	71	69
	Bucharest and its county seats, except Tulcea	319	318	324	339
3 stars	**Total**	**6880**	**7193**	**7658**	**8350**
	Spa resorts	459	519	581	578
	Bucharest and its county seats, except Tulcea	1236	1237	1284	1407
2 stars	**Total**	**2747**	**2846**	**2810**	**2858**
	Spa resorts	195	210	187	188
	Bucharest and its county seats, except Tulcea	330	328	334	319
1 star	**Total**	**679**	**684**	**619**	**747**
	Spa resorts	81	79	29	84
	Bucharest and its county seats, except Tulcea	58	57	58	64
Not classified by stars	**Total**	**92**	**88**	**141**	**87**
	Spa resorts	10	8	59	10
	Bucharest and its county seats, except Tulcea	14	16	19	15

Source: Author's adaptation according to data from the Insse.ro (2020)

Approximately 17% is the share of accommodation units with 4 and 5 stars/flowers, respectively, in the county capitals, while in spa resorts, the share is approximately 8%, on average, for the period 2021–2024, in total units. We can observe an increasing trend of tourist services in urban areas, thus counterbalancing the natural advantage held by spa resorts (Fig. 9.2).

The number of accommodation units classified with 1 and 2 stars/flowers registered increases of up to 8% in spa destinations, in contrast to urban areas, where a decrease of up to 25% was observed. These developments suggest a low demand from tourists for establishments of this type, in the urban area, with the emphasis being placed on the quality of services.

Accommodation units classified with 3 stars/flowers are the most numerous, with a proportion of 61.68% in total units, in 2024, followed by those classified with

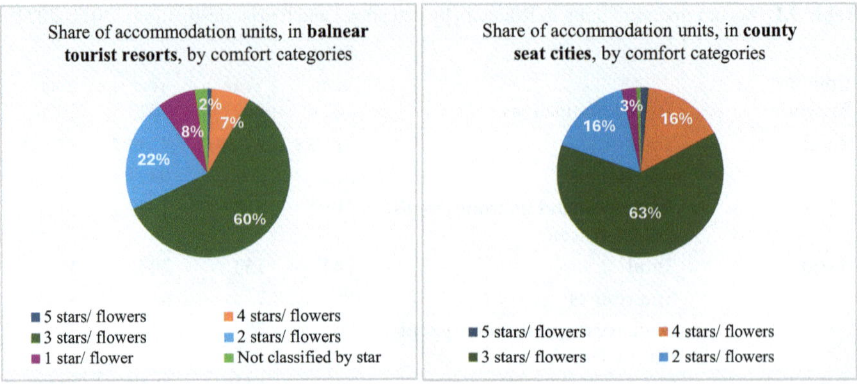

Fig. 9.2 Share of accommodation units, by comfort categories, 2021–2024. (Source: Author's adaptation according to data from the insse.ro (2020))

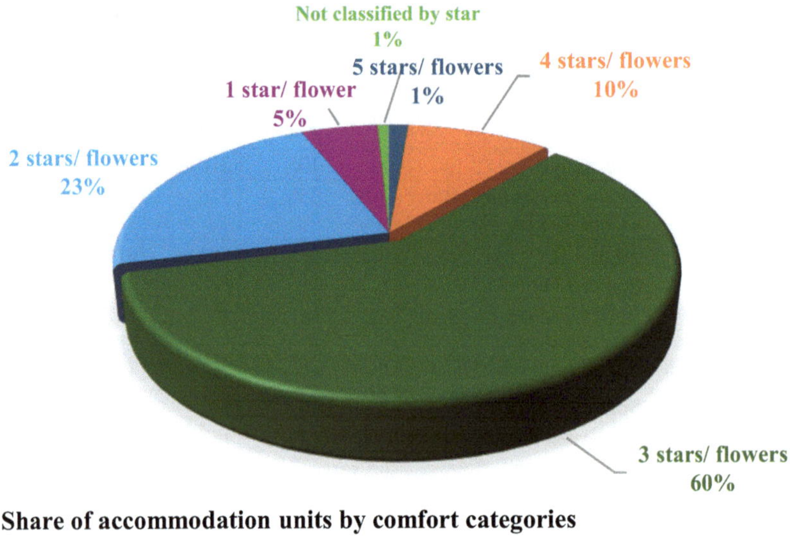

Fig. 9.3 Share of accommodation units, by comfort categories, 2021–2024. (Source: Author's adaptation according to data from the insse.ro (2020))

2 stars/flowers, 2858 units. On average, between 2021 and 2024, at the national level, the largest share is of 3-star accommodation units—about 60%, followed by those classified with 2 stars—22% and those with 4 stars—10% (Fig. 9.3).

Another indicator considered for the research is **the number of** arrivals in tourist reception structures, between 2021 and 2024, at the national level, considering the following:

- Between 3.55% and 4.23% are tourist arrivals in 5-star accommodation units.
- Between 27.32% and 29.76%, there are tourist arrivals in 4-star accommodation units.

- Between 45.44% and 45.92%, there are tourist arrivals in 3-star accommodation units.
- Between 9.20% and 10.36%, there are tourist arrivals in 2-star accommodation units.

Most tourist arrivals are at accommodation units classified with 3 stars, one of the reasons being the quality–price ratio, but there is also an annual increase in tourist preferences for accommodation units classified with 4 stars, increasing annually, the number of tourist arrivals, on average by 1.22%.

This increase in interest in comfort and the superior quality of services offered in tourism can also be seen in the decrease in the number of arrivals in accommodation units classified with 2 stars, reaching 9.2% of the total arrivals nationwide (Fig. 9.4).

Occupancy is another benchmark in tourism, measuring the efficiency of each accommodation unit and the attractiveness of a destination, and is also used in making future decisions.

Following the query of the INS –Tempo Online database, we found, according to Table 9.2, an occupancy rate of 38.8% and 42.5%, respectively, for 4- and 5-star hotels, while 3-star hotels 37.6%. And in the case of tourist pensions, the situation is similar, with those with a 4- and 5-star classification having a higher occupancy rate than the 3-star ones. Comparing the accommodation units classified with 1.2 stars with those classified with 4.5 stars, we can observe an occupancy rate of less than half, which registers an occupancy rate that does not exceed 13.9%.

Considering the analysed data, we can say that the performance of an accommodation unit is influenced by the comfort category.

To find a link between the classification and the performance indicators for tourist units, we analysed the accommodation units in Bacău County, respectively, those in the municipality of Bacău, compared to the units in the Slănic Moldova spa resort.

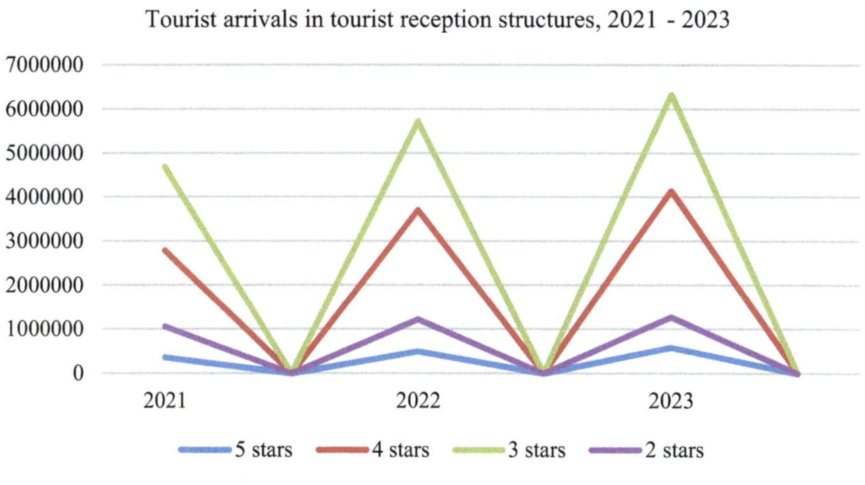

Fig. 9.4 Tourist arrivals in tourist reception structures, 2021–2023. (Source: Author's adaptation according to data from the insse.ro (2020))

Table 9.2 Tourist accommodation capacity net utilization index, 2021–2023

		Periods		
		Year 2021	Year 2022	Year 2023
Types of tourist reception structures	Comfort categories	Percentage	Percentage	Percentage
Hotels	**Total**	32	37.7	39.2
	5 stars	27.6	38.2	38.8
	4 stars	31.7	40.3	42.5
	3 stars	33.1	36.7	376
Hostels	**Total**	18.3	19.9	20.7
	5 stars	32.2	29.8	29.8
	4 stars	23.6	25.9	27.7
	3 stars	18.5	20.4	21
	2 stars	13.1	13.5	13.9

Source: Author's adaptation according to data from the insse.ro (2020)

Table 9.3 Selection criteria for the sample

		Slănic Moldova	Municipality Bacău		
Star categories	Unit type	Not. of units	Not. of accommodation	Not. of units	Not. of accommodation
5	Villa/ Guesthouse	1	8	1	6
4	Tourist guesthouse	1	15	1	10
	Hotel	1	89	1	96
3	Tourist guesthouse	1	10	1	14
	Rooms for rent	1	7	1	10
	Hotel	1	43	1	49
	Tourist villa	1	21	1	6
2	Rooms for rent	1	3	1	2
1	Tourist guesthouse	1	7	1	4

Source: Author's adaptation according to data from the Ministry of Economy, Entrepreneurship and Tourism (2020)

According to the data published on the website of the Ministry of Tourism, 79 accommodation units are registered in the municipality of Bacău, a number close to 72 units registered in the Slănic Moldova spa resort. For the analysis of the performance indicators of the companies with tourist activity, seven accommodation units were selected from the two localities (seven units from Bacău municipality and seven units from Slănic Moldova), having similar selection criteria (type of unit, category, and number of accommodation spaces), according to Table 9.3. Similar selection criteria were used for the two localities, so that the results obtained from the analysis conducted could be comparable.

For the selected accommodation units, performance indicators such as **turnover, profit, and profitability were analysed** (Table 9.4).

Table 9.4 Performance indicators for accommodation units in Bacău and Slănic Moldova

				2023				
	Stars	Code	Accommodation	Turnover	Total costs	Profit	Return on revenue	Return on spend
Slănic Moldova	5	22351770	Siam	277,705	256,874	20,831	7.50%	8.11%
	4	17198078	Crystal	2,902,448	1,875,844	1,033,730	35.62%	55.11%
	4	16774512	Pearl	10,655,854	9,509,901	3,335,078	31.30%	35.07%
	3	945187	White House	75,771	93,812	−18,041		
	3	38580711	Casa Andra	236,860	213,808	27,237	11.50%	12.74%
	3	29879987	Hotel Dobra	8,028,838	8,318,029	1,715,068	21.36%	20.62%
	3	969739	Villa Ioana	2,437,014	984,845	1,428,486	58.62%	145.05%
	2	43032170	Ancuta	188,182	199,754	−13,172		
	1	2477520	Maria Guesthouse	26,220	30,158	−3938		
Municipality of Bacău	5	32169220	Charlote Boutique	537,996	512,658	25,338	4.71%	4.94%
	4	951138	Hotel Dumbrava	21,668,615	14,740,535	6,928,080	31.97%	47.00%
	4	45927296	Casa Sturza	486,386	353,848	132,538	27.25%	37.46%
	3	946743	Hotel Bistrița	22,523,830	18,791,272	3,732,558	16.57%	19.86%
	3	28019909	Daciana Guesthouse	1,209,784	1,109,202	100,582	8.31%	9.07%
	3	18452211	Vila George Apostu	67,875,350	60,487,277	7,388,073	10.88%	12.21%
	3	6673413	Vila Royal	2,389,732	2,183,704	206,028	8.62%	9.43%
	2	28501796	Hostel Holland	724,725	837,578	−112,853		
	1	31963353	Carmen Rooms	32,824	34,543	−1719		

Source: Authors' processing from the database of the Ministry of Economy, Entrepreneurship and Tourism (2020)

The analysis of the data presented in Table 9.4 indicates that there are no significant differences either in terms of the performance indicators of the accommodation units in the spa resort of Slănic Moldova and the municipality of Bacău, nor in terms of their number, with 79 accommodation units being registered in Bacău and 70 in Slănic Moldova.

Following the analysis conducted, it was found that accommodation units classified with 1 star or 2 stars recorded a loss in 2023, while units classified with 4 and 5 stars recorded profit, with values between 132,538 and 6,928,080 (Fig. 9.5). The profitability indicator is a performance indicator that shows us the success of a company.

Thus, analysing in parallel the selected units, according to the criteria in Table 9.3, we can conclude that success is on the side of companies rated with a higher number of stars. The higher the quality of the services offered, the more tourists are attracted

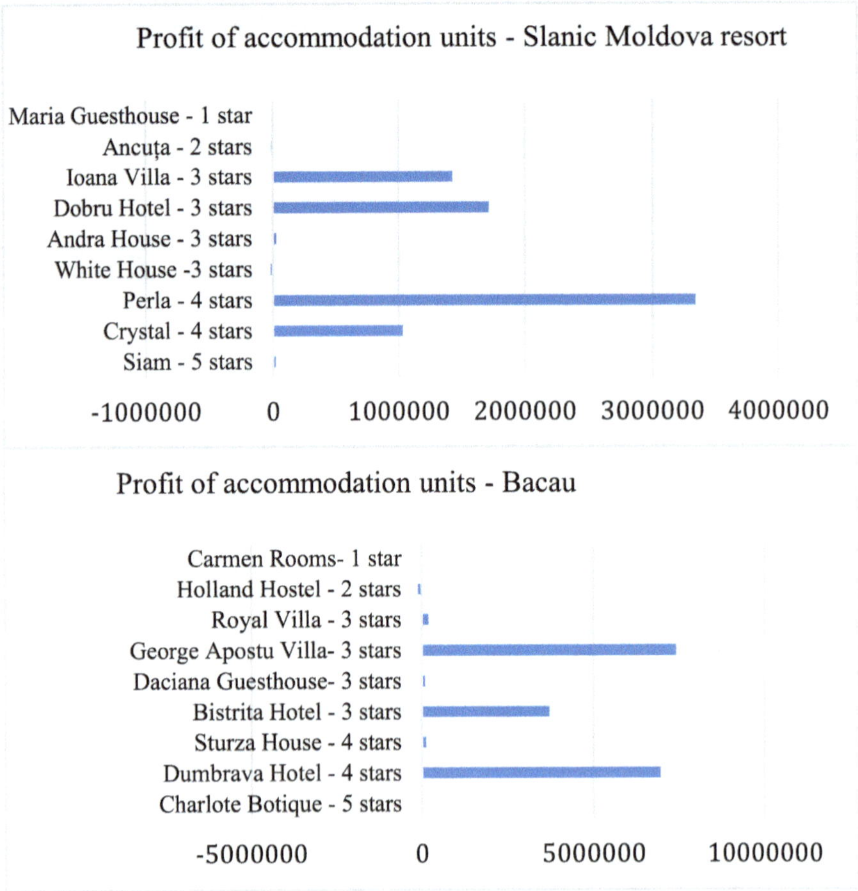

Fig. 9.5 Profit of accommodation units, 2023. (Source: Author's adaptation according to data from the Ministry of Economy, Entrepreneurship and Tourism (2020))

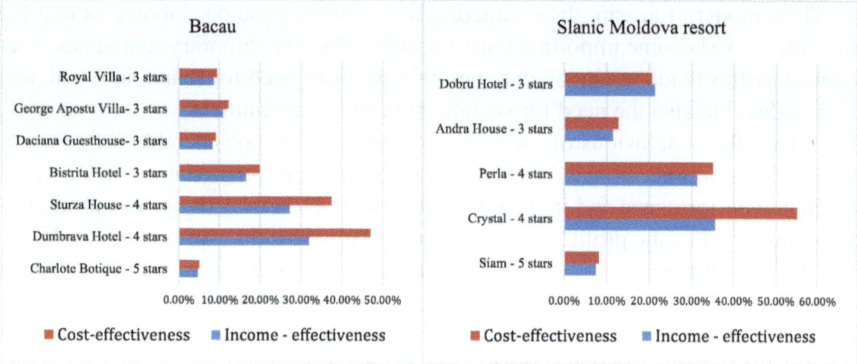

Fig. 9.6 Profitability of units—2023. (Source: Author's adaptation according to data from the Ministry of Economy, Entrepreneurship and Tourism (2020))

to that accommodation unit, which implicitly leads to the profitability of the travel company.

Another performance indicator analysed is profitability, which is an economic and financial indicator that measures a company's ability to generate profit in relation to the resources used.

The cost-effectiveness calculated for the sampled enterprises shows a higher or lower efficiency of resource use. In parallel, an analysis of the profitability of sales was also conducted, calculated by correlating profit with revenues, to obtain a more accurate picture of the performance recorded by the analysed units.

The profitability calculated by the two methods, the profit/income ratio and the profit/expense ratio, for the 4- and 5-star tourist units in the sample, is between 31.3% and 55.11%, while for the units classified with 3 stars, it is between 10.88% and 20.62%. This result highlights a higher level of profitability among accommodation units with a higher degree of classification. For 2- and 1-star units, profitability was not calculated because they recorded a loss (Fig. 9.6).

9.5 Conclusions

In the history of tourism, the rule has emerged according to which tourist resorts are perceived as much more attractive than other established locations.

The analysis of the data from this chapter indicates that there are no significant differences either in terms of the performance indicators of the accommodation units in the spa resort of Slănic Moldova and the municipality of Bacău, nor in terms of their number, with 79 accommodation units being registered in Bacău and 70 in Slănic Moldova. Moreover, at the national level, the number of accommodation units in the county capital cities is much higher than the number of accommodation units in spa resorts.

Thus, modern tourism, the competing link of the classic one, shows us that the big cities have become important tourist centres. The contemporary tourist has other characteristics than the classic one, not only people's need for relaxation, rest, and knowledge, but also the need for socialization and entertainment.

In fact, the conclusions of the research underline the considerable influence of the level of comfort, reflected by the classification of accommodation units, on the economic performance and indicators specific to the tourist activity. The profitability, determined by the profit/income and profit/expense ratio, varies between 31.3% and 55.11% in the case of 4- and 5-star units, while for 3-star units, it is between 10.88% and 20.62%. These differences indicate a higher profitability among units with a higher degree of classification, confirming the positive correlation between the desired comfort and the economic and financial results.

The study concludes that success is on the side of companies rated with a higher number of stars. The higher the quality of the services offered, the more tourists are attracted to that accommodation unit, and implicitly, this aspect leads to the profitability of the travel company.

Regarding the influence of location on accommodation units, we can say that non-established localities can become viable competitors, due to factors such as lower rates, increased accessibility, the organization of festivals, and the existence of an infrastructure favourable to shopping, educational, as well as business tourism.

References

Basit, A., Khan, A. A., & Qazi, T. F. (2021). Who played more deserves to be seen? Country-level ranking and classification based on travel and tourism performance. *SJESR, 4*(2), 119–130.

Boghean, C. (2007). Perspectives for the sustainable development of Romanian tourism. *Journal of Studies and Research in Tourism, 3*, 58–62.

Chele, A. C., Podaru, G., & Stratila, V. (2024). The role of special events in tourism. *Cactus-The Tourism Journal for Research, Education, Culture, and Soul, 6*(2), 93–106.

Ciobanu, V. (2022). Quality – A determining factor of sustainable tourism. *International Relations. Plus, 22*(2), 145–152.

Corluka, G., Mikinac, K., & Milenkovska, A. (2016). Classification of the tourist season in coastal tourism. *UTMS Journal of Economics, 7*(1), 71–83.

Costea, M., Hapenciuc, C. V., & Arionesei, G. (2016, September). Romania versus Bulgaria: A brief analysis of the competitiveness of seaside tourism. In *The Proceedings of the CBU International Conference* (Vol. 4, pp. 471–482).

Győri, M. M. (2011). Contemporary development of tourism in Chile. *Studia Universitatis Vasile Goldis, Arad-Economic Sciences Series, 21*(1), 491–500.

Hapenciuc, V., Stanciu, P., Condratov, I., & Nistoreanu, P. (2009). Safe tourism in crisis situations, appreciations in the case of Romania. *Journal of Studies and Research in Tourism, 7*(7), 47–56.

Insse.ro. (2020). *TIME Online*. [Online] http://statistici.insse.ro:8077/tempo-online/#/pages/tables/insse-table. Accessed 6 Feb 2025.

Just.ro. (2025). *NORM 15/04/2010 – Legislative portal*. [Online] https://legislatie.just.ro/Public/DetaliiDocument/119050. Accessed 8 Feb 2025.

Ministry of Economy, Entrepreneurship and Tourism. (2020). *Ministry of Economy, Entrepreneurship and Tourism*. [Online] https://turism.gov.ro/web/. Accessed 15 Mar 2025.

Mulet-Forteza, C., Ferrer-Rosell, B., Cunill, O. M., & Linares-Mustarós, S. (2024). The role of expansion strategies and operational attributes on hotel performance: A compositional approach. *arXiv preprint arXiv:2411.04640.*

Suslenco, A., & Movila, I. (2018). Possibilities for developing and capitalizing on the tourism potential. In *Ensuring the economic-managerial viability for the sustainable development of the regional economy in the conditions of EU integration* (pp. 55–59). Universitatea de Stat „Alecu Russo" din Bălți.

Verstiuc, R. (2022). Creating offers for 4-star hotels in the Republic of Moldova. In *The scientific annals of the State University of Moldova* (pp. 157–159). Moldova State University / Universitatea de Stat din Moldova.

Chapter 10
Curating the Countryside: Emerging Business Models in Cultural Tourism in Rural Areas

Greg Richards

Abstract COVID-19 underlined the potential of rural areas for cultural tourism as urban dwellers flocked to the relative isolation of the countryside. Following the pandemic, the expected growth of rural tourism failed to emerge, leading to a reassessment of rural tourism strategies. As localities seek to increase their attractiveness for tourism, the basic natural resources of rural areas are being transformed and enhanced through a number of new developments, including new types of cultural tourism. These emerging cultural tourism business models are being selected and disseminated through expert curation, algorithmic curation, and social curation. Based on recent literature and drawing on data from the Crocus Project, this chapter aims to highlight the influence of curational strategies on the development of rural cultural tourism. Our analysis identifies a wide range of business models, including value propositions centred on the provision of accommodation, the staging of experiences, co-creating culture, and mobility. Curated "gastroscapes" are particularly prominent in the packaging of the rural area for tourist consumption, and festivals and events also provide a means of making rural culture accessible to tourists. Digitalisation is also ushering in a host of platforms that algorithmically and socially curate the rural to make it more accessible and readable for outsiders. The main conclusions are that rural areas are gradually being colonised by formerly urban models of curation, in which the consumption of the rural is aligned to the tastes of the urban dwellers who constitute the main audience for rural tourism.

Keywords Cultural tourism · Creative tourism · Curation · Rural areas · Business models

G. Richards (✉)
Tilburg University, Tilburg, The Netherlands
e-mail: g.w.richards@tilburguniversity.edu

A. L. Negrusa, M. M. Coroş (eds.), *Innovation, Sustainability, and Growth in a VUCA Environment*, Springer Proceedings in Business and Economics,
https://doi.org/10.1007/978-3-032-11639-0_10

10.1 Introduction

Rural areas have tended to be on the margins of debate about cultural tourism growth. The bulk of cultural tourists flock to crowded city centres full of tangible heritage sites, while most rural tourists are in search of the peace and tranquillity of the countryside. Rural areas generally lack the "must see sights" that dominate the agendas of the urban cultural tourist. Instead, they often form a backdrop to a range of leisure activities. Rural areas have also been relatively ignored in development agendas, with urban transformation and challenges tending to attract most attention. This is not surprising given the growing proportion of people living in cities, currently about 80% of the global population.

But as Rem Koolhaas pointed out in his curation of the *Countryside, The Future* exhibition at the Guggenheim New York, cities still only account for 2% of the total land area, whereas the rural 98% gets little attention. One reason is that cities are the focus of the global "space of flows" that dominates the economy, society, and culture (Castells, 2000). In contrast, the countryside is the realm of the "space of places," localities disconnected from financial markets and global media, dedicated to feeding the cities and providing recreation opportunities for their inhabitants. This has increased the marginalisation of rural places, which are seen as peripheral to the urban, not just in a spatial sense but also socially and culturally.

The Koolhaas exhibition was staged in New York in 2020, during the COVID pandemic. This was also a moment at which it seemed there would be a retreat from the cities and a renaissance for the countryside. But, in fact, the opposite seems to have happened. Major cities are once again buzzing with tourists, and real estate prices in cities are soaring. Rural areas seem to have returned to their previous status as appendages of the urban world.

But rather than being simply forgotten, as Koolhaas suggested, the rural is also being rediscovered and reinvented. I want to suggest here that what we are seeing is a new post-pandemic form of colonisation of the rural by the urban world. In the past, this process was most evident at the margins of cities that ate up their rural surroundings with suburbs and warehouses, as an extension of the productive spaces of the city. Now we are increasingly seeing the rural colonised by consumption practices formerly connected with cities, such as the development of flagship attractions, idealabs, smart tourism, gamification, street art, and the creative industries. The organisation of these practices is also adopting urban forms, including platformisation and curation. We are now witnessing the growth of "overspill tourism" based on these "new cathedrals of (rural) consumption" (to paraphrase Ritzer, 1999).

This conceptual paper examines the dynamics of this countryside colonisation and analyses its effects on cultural and creative tourism (CCT) in rural areas. It draws extensively on the Crocus Project (2025), which is currently examining the development of new business models in this area. This chapter examines recent studies on CCT in rural areas, paying particular attention to the changes the countryside is currently facing, the development of CCT in rural areas, and the emergence of new models of rural CCT. In particular, we will highlight the importance of intangible cultural heritage (ICH) in these developments, which arguably represents a major resource for future rural development.

10.2 Literature Review: Rural Change, Tourism, and Culture

Rural change is nothing new. The trend towards open-air museums at the end of the nineteenth century was a reaction to the disappearance of rural livelihoods and lifestyles through industrialisation and modernisation (Shafernich, 1994). As cities expanded, rural areas faced challenges of population decline and aging, loss of traditional economic activities, and lack of accessibility relative to cities.

The globalisation ushered in during the twentieth century also brought new types of change. Tourism was central to this because it helped to consolidate the transformation of rural areas from productive to consumptive spaces. Instead of agricultural production being the predominant source of income, tourism and other services became increasingly important. By the beginning of the twentieth century, rural areas faced further challenges from the energy transition and climate change. Rural dwellers have lower income per capita, a higher percentage of the population is at risk of poverty and social exclusion, lacking access to basic infrastructure and services, and with lower levels of access to fast broadband internet.

Rural areas are currently being transformed to meet the needs of urban populations, both in terms of production and consumption and, increasingly, in terms of the societal challenges and climate change. The few farmers who remain are now being asked to limit their productive activities to meet carbon emissions targets. Rural land is being given over to solar power generation or biofuels.

These challenges have given rise to a search for new meanings and uses for rural places. In 2021, the European Commission published a vision for rural areas of the European Union (European Commission, 2025), which account for 45% of the EU land area. This includes a "rural pact" to bring stakeholders together to meet the needs of rural communities and develop an EU rural action plan. Important elements of this new approach include supporting the economic diversification of rural areas, utilising local strengths, knowledge, and opportunities.

Tourism, and particularly CCT, has been identified as an important tool for diversification. Although CCTs have been seen as major drivers of urban-based tourism in recent decades, they have been less emphasised in rural areas, where enjoying nature and escaping the stress of urban life have often been seen as the major drivers of rural tourism. In addition, rural areas often lack the tangible heritage resources that are common in urban areas.

As styles of CCT evolve, however, this a-cultural view of the rural is beginning to change. Rural areas are increasingly seen as places where everyday culture and creativity can be experienced by visitors seeking authenticity and engagement. In particular, creative tourism has been developed in many rural areas, providing active involvement of tourists in the creativity of rural areas (Cloke, 2007). This transformation is being stimulated not just by demand from tourists but also by a creative turn in rural areas, linked to the emergence of a rural "creative class" and the identification of the creative industries as a potential source of rural growth (Herslund, 2012). Many clustering initiatives have been researched in rural areas, including creative hubs (Hill et al., 2021), artists' residencies (Almeida et al., 2025), and "creative outposts" (Brouder, 2019). These clusters of creatives mimic, to some extent, the creative buzz of cities and provide opportunities to articulate with the market. In particular, these clusters are driven by rural tourism, with purchases by visitors keeping many rural craftspeople afloat. Research is emerging on creative hotspots in the countryside (Richards et al., 2024), but rural creatives tend to be attracted by tranquillity rather than creative buzz. Having relocated to rural areas, creatives then seek each other out for mutual support and ideas, but this has little to do with their location choice (Leick et al., 2023).

This research indicates that new arrivals in the countryside become drivers not just of new consumption opportunities but also new modes of production.

Many of these economic transformations in rural areas are also related to social changes, including the in-migration of new groups, including tourists and migrants. These economic, social, and cultural changes are transforming the relationship between rural heritage, rural populations, and visitors. A number of commentators have pointed to emerging contradictions between the modernisation of rural life and the continued desire of many visitors and migrants to maintain the "authenticity" of rural culture. For example, incomers are often keener to preserve the "unspoilt" rural landscapes that attracted them to relocate than the original rural inhabitants themselves, who tend to prioritise economic development (Boucher et al., 2017). There is a danger that the ICH and lifestyle of rural areas become "musemified" and packaged for tourist consumption (Perkins, 2006).

Culture, and in particular ICH, has become a key resource for sustainable development in the countryside. At the same time, it has become a resource contested by many different stakeholders who emphasise different aspects of ICH as important to their needs. The following section examines the developing debate on the role of rural culture in tourism.

10.3 Rural Tourism and Culture

Rural areas have long been important tourism destinations, providing recreational opportunities for the growing number of urban dwellers. In recent decades, there has also been a substantial growth in "residential tourism" in rural areas, with the purchase of second homes in areas close to major cities being an important factor (Stiman, 2020).

Rural tourism has largely been framed as a form of escape from the urban, with city dwellers seeking peace and tranquillity as well as typical rural activities characterised by being close to nature, and based on "traditional" ways of life. Rural tourism demand has been stimulated by higher levels of education, growing interest in heritage and tradition, and a search for authenticity and personal contact in a globalising world (Long & Lane, 2000). The rural world, therefore, has become increasingly enlisted into the "experience economy" (Pine & Gilmore, 1999), with experiences of both nature and rural culture forming an important basis of the tourism product.

The search for authenticity and personal contact can, however, produce friction. Large numbers of tourists flooding rural areas (including waves of city-based influencers!) can create antagonism among residents as their daily lives are disrupted by onlookers. Outsiders also penetrate and shape rural life by utilising local facilities, including housing and amenities, sometimes limiting access for locals. There have been increasing debates about the affordability of rural housing, with the growth of tourism seen as a cause of rising prices in many areas.

Changing the current, often adversarial, relationship between visitors and locals is one key to boosting local economies and local culture. Positive encounters between these groups can be a means of generating new ideas and injecting resources. The Crocus Project has adopted the view that such processes can best be stimulated by taking a place-based approach to the development of CCT in rural areas (Richards et al., 2024). This is also a cornerstone of the creative tourism concept (Richards & Raymond, 2000), which, over the past quarter century, has developed in new and sometimes surprising directions. Place resources and people become the catalysts for creative activities that provide alternatives to traditional, more passive forms of cultural tourism. The ways in which people relate to place are deeply culturally embedded, and place attachment is a strong creative impulse for locals that can also forge bonds with visitors.

The ways in which a place-based approach can support rural communities are examined in Norway by Clemetsen, Stokke, Barane, and Haraldseid (2021), who focus on nature-based tourism. They argue that the role of culture and creativity in rural areas may often be underestimated, because tourism is often seen as based on activities such as hiking, cycling, kiting, or angling. However, these activities take place in a local community possessing skills, knowledge, and experiences that are related to these practices and which are of interest to the visitor. Visitors also contribute their own skills, knowledge, and experiences to the practice, making them open to local culture, as well as enriching local culture and practices. They underline the reciprocity between visitors and locals in nature-based tourism, which often has important cultural elements. The development of such practices also involves the history of the practice itself, which, in turn, becomes an element of local culture as well as the knowledge and skills contributed by outsiders.

For Clemetsen et al. (2021), tourists do not visit a destination: they visit a place and experience and contribute to the creativity and culture of that place, as well as the practices it is linked to. Place-based heritage and cultural practices, therefore, become important in creating distinction between rural places and in generating links between the rural inhabitants and visitors. This more holistic approach to the rural means that the countryside becomes more than a location for 'doing' rural activities: it becomes a setting to be experienced through the landscapes, people, and culture that are intertwined in rural places. This is becoming increasingly evident in the ways in which rural places are produced and marketed for tourism.

10.4 The Curation of Rural Places

We are currently seeing the "platformisation of consumption" (Śledziewska & Włoch, 2021), which is most obvious in cities, where high-density consumption helps to support major platforms such as Uber and Airbnb. Platformisation entails the growth of platform-driven, algorithmically managed, and data-centric systems into previously non-digital economic functions. Tourism is one area becoming increasingly platformised, with offers of city tourism being a major focus. Pre-pandemic, Capineri and Romano (2021) reported 50,000 Airbnb Experiences in 1000 cities worldwide. Although Airbnb suspended its Experiences Programme shortly afterwards, it bounced back strongly post-pandemic, and there were more than 100,000 experiences by July 2024. In May 2025, for example, 392 experiences were listed in Tuscany, Italy, many of which are in rural areas, such as truffle hunting, wine tasting, organic farm visits, horse riding, and cooking classes.

The growth of the platform economy is now spreading to less populated rural areas as well. Even though the level of consumption is lower, platforms can link rural experiences to urban markets, helping to boost demand. We are now also seeing the development of rural-focused platforms offering a wide range of experiences in the countryside. Richards (2024) has identified a "curational turn" in tourism, which is related to the growing "choice stress" produced by the flood of consumption opportunities in the experience economy. There is a growing range of curators who are helping consumers to make choices by selecting and recommending things to experience and places to be.

Curation is defined as "the meaningful and persistent selection and organisation of content and experiences to relate production and consumption and facilitate better choices to increase aesthetic value" (Richards, 2024). Curation is therefore related to stylistic innovation, which creates high symbolic value, which is negotiated between aesthetic selectors.

Richards (2024) identifies different forms of curation in tourism, including expert curation, social curation, and algorithmic curation:

1. Expert curation is undertaken by an expert or a group of experts who curate content for larger audiences. These experts include the traditional museum curator or travel blog writers.
2. Social curation is based on a large group of people who collectively find, organise, and share content (e.g., Flickr). This can also be termed "crowdsourcing," most frequently seen in tourism on websites such as Tripadvisor.
3. Algorithmic curation is a process whereby a computer program selects content to present to users based on algorithms encoded by human programmers, such as

> Netflix or Google search. In tourism, Airbnb uses algorithmic curation to present selections of accommodation to consumers.

Richards (2024) also identifies a growing market in "place curation," in which Destination Marketing Organisations (DMOs) and other intermediaries curate the experiences in their places to appeal to particular audiences. One of the interesting effects of place curation is that it facilitates the development of platform economy approaches to tourism in rural areas in similar ways to urban areas. We already see this, to some extent, in the activities of Airbnb and similar travel-related platforms, which offer "hyper-local content curation" (Divakaran, 2017) by drawing on user-generated content and assembling stories relating to engaging places and people. This is a means of categorising the experience through curation in order to support the consumer choice process.

This is a strategy being developed through many different rural-based platforms, which often centre on place as the central value proposition. The place has become a holistic bundle of rural experiences involving the physical environment, the people, and experiences. This holistic approach to place requires a certain level of curation.

As Explore Balkans says: "Pick one or more of our 8 carefully curated tours and immerse in exploring crystal clear lakes and rivers, mountaintop villages, hiking breath-taking areas, ski-shoeing and much more!"

Gravel Travel sees curated travel as "a more refined and personalized way of planning … an experience that you won't soon forget. Travel curators are the brains behind the operation. They partner with a select group of hotels, restaurants, and special attractions that offer something unique, like hitching a hot air balloon ride over the temples of Myanmar." Curated travel is different because:

> Rather than crossing off bucket-list destinations simply by showing up and moving on, a curated trip offers a more immersive experience and handles all of the logistics for you. That might mean skipping the long lines at tourist traps and getting VIP access to exclusive events and experiences.

For Ciao Puglia Travel (2025), "your gateway to curated travel experiences in southern Italy", curation is about answering the questions of travellers based on expertise and experience: "Wondering where to go? Where to stay? What to do? What to eat? We will answer these questions and any others you have to create the vacation of your dreams." There are now many curated travel companies offering similar choice support services in rural areas. But more significant is the growth of travel platforms specialised in rural experiences.

One recent arrival is BeCountry (2025) from Italy, which prompts farmers to "Upload an experience and promote your local business: Offer an authentic experience at your agricultural estate and share unforgettable moments with the BeCountry community. Start telling your story today."

BeCountry is big on passion: passion "for good food, for authentic experiences, for the regions of Italy and the excellent products that they offer. We are a platform at the service of users who share this passion. We believe that farms are the beating heart of the Italian countryside, contributing work handed down across generations."

BeCountry seeks to create distinctiveness by crafting traditions and culture into stories that can be shared with visitors. The BeCountry mission is to:

- Strengthen Italian communities
- Encourage local tourism
- Give a voice to farms

It also emphasises sustainability, with "zero kilometre experiences" related to local products and their production, transformation, or conservation processes, which basically relates to creative tourism.

For example, the "shepherd's day with cheesemaker's workshop" allows guests to:

> participate in every step of milk processing, discovering the value of animal care and dairy processing …. After collecting the milk, you will enter our dairy workshop, where we will guide you in creating your own goat cheese using traditional methods passed down through generations. To end on a high note, you will participate in a tasting of 0-mile products, with a selection of fresh and aged cheeses, artisanal cured meats, and freshly prepared hot ricotta.

The activities include the following:

- Hand milking the goats, to learn an ancient art and connect with the animals
- Guided tour of the farm, to learn about farm life and connection to the land
- Cheese-making workshop, with explanation of traditional techniques
- Active participation in milk processing, from curd to final processing
- Tasting of artisanal cheeses, typical cured meats, Cassio bread, and freshly prepared warm ricotta
- Opportunity to take home your own cheese, for a tasty and unique souvenir

This experience costs 80 euros per person through the platform, which also enables you to give the experience as a gift to somebody else.

The experiences are curated into categories such as "experience gifts: four original ideas for lovers of nature and authentic flavours" or "Experiences to give to a couple: Five authentic ideas worth more than an object" or "Under the stars, among the rows: vineyard dinners in Lombardy."

With the platformisation of the rural world, the countryside is increasingly subject to curatorial influences and overspill tourism from cities. Posts by influencers can generate crowds in relatively remote locations overnight. Specific places are singled out for consumption by particular groups of urbanites, such as digital nomads. As digital nomads move to rural places for cheap rents and cultural experiences, they bring with them urban consumption patterns and a demand for fast Internet access.

Growing digital nomadism is argued to be especially important for small villages in remote regions, encouraging slow tourism and preserving the authenticity of the destination. Lonardi, Buffa, and Martini (2024) studied the Valsugana valley in Northern Italy, where Impact Hub Trentino has developed coworking spaces in formerly abandoned buildings. Research with stakeholders shows that digital nomad tourism is seen as developing in harmony with the needs of local residents. Digital nomads can also be attracted to the mountains in the low season, and they are perceived to be interested in local traditions and wanting authentic contact with locals.

10.5 New Business Models

The rise of digital nomad tourism is one example of how attempts by rural dwellers to develop, curate, and market experiences to tourists are giving rise to new business models in CCT, or ways in which "various organizations, including public bodies, NGOs, and networks, create value for users and communities" (James et al., 2024, p. 5).

During the Crocus literature review process (Richards et al., 2025), almost 100 different business models were identified in relation to CCT in RRA. Among the most commonly mentioned in the literature were tangible heritage resources (such as monuments and museums), gastronomy, events, and cultural routes. These new business models can be characterised as value propositions relating to moving through, dwelling in, and experiencing rural areas, and co-creating rural culture with locals (Table 10.1).

By far the majority of business models in rural areas relate to experiencing the rural or aspects of rural culture. Gastronomy is an area that has seen significant growth in recent years, driven by the development of regional and local gastronomies and the development of entire "gastroscapes" devoted to different forms of food and gastronomic consumption (Richards, 2021a). For example, New Nordic

Table 10.1 A categorisation of rural CCT business models

Category	Examples	Sources
Dwelling	Accommodation models: second homes, glamping, re-converted heritage	Droli (2019)
Co-creating	Creative tourism and know-how tourism	Richards (2020)
Experiencing	Arts, crafts, literature, and languages	Horjan (2011)
	Battlefield tourism/war heritage	Noivo et al. (2022)
	Events, festivals, and performance	Paniagua (2016)
	Gastronomy	Gyimóthy (2017)
	Luxury tourism and mindful luxury	Leban et al. (2024)
	New technologies	Schuhbauer and Hausmann (2022)
	Religious tourism, spiritual tourism, and Wellness	Trono and Castronuovo (2021).
	Storytelling, mythology, re-enactment, and living history	Korez-Vide (2017)
	Tangible heritage	Tamborrino (2020)
Moving to and through	Cultural routes, biking, walking, Transhumance, and migration heritage ("roots tourism")	Khovanova-Rubicondo (2012)

Source: Crocus Project

Cuisine has been developed in many rural areas of Scandinavia. This culinary movement has a cultural base, having the aim to "promote Nordic products and the variety of Nordic producers – and to spread the word about their underlying cultures" (Leer, 2016). The Nordic cuisine movement also brings together a wide range of stakeholders, "To join forces with consumer representatives, other cooking craftsmen, agriculture, fishing, food, retail and wholesale industries, researchers, teachers, politicians and authorities on this project for the benefit and advantage of everyone in the Nordic countries." This can support new governance models to link regions and sectors, as well as stimulate co-creation between local producers and visitors.

New Nordic Cuisine also features seasonal produce, helping to turn seasonality from a weakness in tourism markets into a potential attractive force. New business models can be innovated to provide a year-round range of experiences. Sauna and wellness tourism, indigenous tourism with the Sámi, and creative experiences related to crafts (Huhmarniemi et al., 2021) are among the many possibilities.

Important gastroscapes can also be found in Southern Europe, as Fernandes and Richards (2021) have illustrated in the case of the Minho region of Northern Portugal. In this region, there is extensive production of local foods on small farms, which are usually served in local restaurants in large portions at low prices, in accordance with long-established cultural practices. This provides the basis for food clusters, such as the *arroz de sarrabulho* served by many of the restaurants in the town of Ponte da Lima. According to tradition, the Moon must be in the first quarter when the pig is slaughtered for this dish. The close link between landscape, farming practices, and local gastronomy makes this a popular destination for regional tourists, who descend in large numbers over the weekend, increasing to several thousands during the *Feiras Novas* in September. The attraction is not just the food or the celebrations, but the attraction of the place, a culture of "Minhoness" as Fernandes and Richards (2021) express it. This regional gastroscape has to be carefully curated through the orchestration of a series of gastronomic and cultural practices. But this is different from cultural preservation or the freezing of local culture. What is important about the Minho is the adaptation of local culture to contemporary influences, now reflected in the gastroscape by the emergence of new innovative restaurants providing new versions of traditional dishes that can reach new target groups.

The rise of new business models in RRA is giving rise to interesting new experiences, often based on co-creation between locals and tourists. The growth of creative tourism in rural areas is supporting the revitalisation of many crafts and traditional activities, such as the creative tourism platform provided by the town of Loulé in Southern Portugal. This offers a range of creative experiences for tourists related to different aspects of local culture, including pottery, palm leaf weaving, wool spinning, wood carving, bread making, and local gastronomy. Creative tourism business models seem particularly useful for rural development, because they can produce direct interaction between local creatives and visitors, as well as supporting local creativity and skills.

Much creative tourism development has been based on traditional crafts embedded in rural communities (Richards, 2021b). However, examples are emerging of new creative practices, such as street art in rural communities. Perkins et al. (2023) report on the rise of "rural murals" as a means of creative placemaking in New Zealand. The painting of giant murals on grain silos in the South Island generated a lot of media coverage, as well as attracting visitors. The murals reflected local stories and identities, with, for example, a giant painting in Ashburton highlighting Hakatere, the pre-colonial Māori name for Ashburton, and the Timaru murals featuring local vegetation and wildlife.

Perkins et al. (2023, p. 10) argue that the murals:

> all take into account and harness available tangible and intangible culture, present these resources in processes of storytelling and curation and highlight the value local people place on creative activity. All the murals have resulted from processes of co-creativity, incorporating multi-party local community participation in various ways.

In other words, such murals, which are usually associated with urban environments, can be curated as effective placemaking tools in rural places too.

Remoteness as Rural Distinction

Remoteness is an aspect of many rural places that is relatively unique. As the world becomes more crowded and busy, the ability to literally "get away from it all" is increasingly becoming difficult. Even a quarter century ago, Long and Lane (2000, p. 302) were already arguing that "rural remoteness (is) so rare that it had become a positive unique selling point for tourism."

Today, remoteness can be seen as a luxury, as Leban et al. (2024) argue in the case of the Faroe Islands. The Faroes market themselves as an exclusive and luxurious destination, with well-preserved cultural heritage and ecological purity. Leban et al. (2024) see luxury experiences as "transformational," rather than the usual view of luxury consumption as wasteful, because "Mindful tourists," free of digital distractions and adapted to the environment arguably, appreciate the value of supporting local businesses and minimising environmental impact. The Faroe Islands experience contrasts with traditional luxury destinations:

> Going to the Faroe Islands showed me what luxury is really about: it's about appreciating the little things, the essential things really of your life. I am thinking nature, people, connections, ancestry, food, culture. You don't get that when you have a hectic lifestyle – you don't have time, you don't have the energy to appreciate these things. (Leban et al., 2024)

Four luxury travel practices were identified by Leban et al. (2024):

- *Hoarding*, or conscious accumulation, cataloguing, and curation of knowledge, with detailed planning
- *Savouring* through time-intensive eating (in luxury restaurants), resting (in luxury hotels), and slow enjoyment of experiences
- *Reconnecting* with nature by disengaging from urbanised and accelerated environments

- *Secluding* or seeking ways to be alone physically and mentally

The Faroe Islands uses these luxury practices to link tourists with locals and the locality, promoting mindfulness, and making luxury consumption more introspective.

The question that might be asked is whether remote places like the Faroes or Greenland might become luxury resort areas for wealthy tourists, while their isolation is preserved through high prices that keep the majority of visitors away. With Donald Trump currently eyeing Greenland for the United States, this is not such a fanciful idea.

10.6 Discussion

As rural areas are increasingly linked to global circuits of tourism consumption, there is increased packaging and curation of experiences. These experiences are generally designed to appeal to urban-based audiences already used to consuming cities through "new urban tourism." The experiences offered on the platforms that might be seen as offering "new rural tourism" often emphasise gastronomy and use events and storytelling as devices to engage tourists. Gastronomy, in particular, is a very common element of rural tourism experiences because of its ability to link together landscapes, people, and intangible culture. The food linked to local culture is usually presented as an "authentic" reflection of local landscapes and agricultural practices, producing a unique blend of preparation and cooking techniques, presentation, and connections with local communities and nature.

Fusté-Forné (2019) examines the "placemaking practice" of mushroom picking in the Pyrenees. He sees wild mushroom gathering as an important seasonal lifestyle practice because it shows how locals give meaning to place. Even though a large number of participants come from cities, many have family links with rural areas via their relatives, and they relive previous trips to the mountains. Mushroom picking is a sustainable rural cultural practice, which reinforces place connections and identity. Such practices are now being integrated into entire foodscapes.

In some ways, the foodscapes of new rural tourism mimic the dynamics of urban gastronomic practices, such as farmers' markets and gatherings of food trucks. There is a wave of new food curators in cities, who are creating gastronomic socialising spaces that promote relaxation and enjoyment and generate festive atmospheres. These features are also increasingly found in rural areas, where gastronomic festivals and other events offer opportunities for socialisation around shared food rituals (Cardoso et al., 2014).

There are also growing signs of place curation beginning to emerge in rural areas, as the development of a "destination marketing organisation for gastronomy" in Greece shows. This organisation will promote rural areas based on their gastronomic offerings and the landscapes and intangible culture that support these. It is important that destinations begin to develop effective governance mechanisms that can support such place curation efforts, because otherwise rural places will increasingly be dominated by algorithmic curation, as we are already seeing in many cities. The spread of Airbnb-like curational strategies to rural areas makes them not only more accessible but also more vulnerable to the "serial reproduction" of their ICH.

In order to counter the effects of platformisation, RRA should perhaps consider moving to place-based curation systems, involving a wide range of stakeholders. The BeCountry platform, based on alliances with local farmers, is perhaps one model that could be developed more widely.

10.7 Conclusions

In terms of the changes and challenges that rural areas are currently facing, our review indicates that the countryside will continue to be colonised by the desires of the urban world. However, the nature of those desires is shifting. On the one hand, the role of the rural as an escape from the pressures of urban is changing, as the focus shifts from tranquillity and rest towards authenticity and contact. Tourists will increasingly want to experience the "real" rural and meet genuine rural people. This will place a premium on the construction of rural authenticity, meeting the needs of visitors. There is a need to shift the role of the rural from destination to place, not letting tourist desires set the agenda, but rather co-creating place with locals.

This will require the development of new modes of rural tourism, which recognise the role of culture and heritage in shaping nature. Essentially, this implies a shift from an activity-based to a place-based approach to rural tourism development. Place-based development offers the opportunity for local producers to work together in developing experiences for tourists, and incorporating many different aspects of local culture into the offer. The requires collaborative rather than competitive approaches to development, something which can be challenging in traditional rural settings. It is important for development initiatives to offer a vision of what new approaches to tourism development can achieve, and how these can benefit local people and local places.

As consumer choice is increasingly mediated by different forms of curation, the production of rural places will also need to change. We can see different phases of "curation of the rural" emerging. In the bucolic phase of rural tourism development, the countryside was an uncomplicated and uncurated space for activities related to nature and rural calm. However, the growth of tourism has increasingly pressured rural spaces, particularly those close to cities. This has seen the development of "spillover curation," with platforms such as Airbnb moving into rural areas close to popular cities, such as Barcelona, or already busy rural areas, such as Tuscany.

What we now seem to be seeing is the development of "curated ruralities," in which dedicated platforms emphasise the authenticity of rural places and communities and also promote the value of tourism to the community and rural culture. In essence, what seems to be emerging is a "new rural tourism"—a mirror of the "new urban tourism" now affecting cities.

The “new urban tourism” has been seen as off the beaten track tourism in cities. In contrast, “New Rural Tourism” is based on seeking authentic rural places, which combine culture and nature, and are usually curated, storied places. The New Rural Tourism is made for the platformed countryside, in which rural experiences can be organised and purchased via the Internet in neat and convenient packages. The advantage for the rural providers of such experiences is that they can harness a ready stream of experienced consumers for a relatively high-priced and standardised product. For the consumer, the advantage lies in being presented with a curated range of rural experiences that bring them into contact with rural people and cultures. Of course, the discovery that the “rural people” providing these experiences may be relatively recent incomers themselves may come as a surprise to some.

In this sense, the platformed countryside mirrors the city, where Törnberg (2022, p. 1) identifies “platform placemaking” involving people “shaping imaginaries of … place in their interests.” In the countryside, we are now seeing similar processes, which will shift our notions of rural places, culture, and tradition. In fact, “countryside” is now one of the 50+ Airbnb Categories that are curated on the Airbnb platform. One might argue that the rural is currently a place being curated in the interests of tourists, rural businesses, and DMOs, all of whom collaborate to facilitate the consumption of the rural.

Acknowledgements This research forms part of the Crocus Project, an EU Horizon Project on Cultural and Creative Tourism in Rural and Remote Areas.

Funded by the European Union	Horizon Europe Grant Agreement No. 101132454. Views and opinions expressed are, however, those of the author(s) only and do not necessarily reflect those of the European Union. Neither the European Union nor the granting authority can be held responsible for them.

References

Almeida, I., Ferreira, C. C., & Silveira, L. (2025). Creative tourism and residents: Interactions in the context of artist residencies in the center region of Portugal. *Investigaciones Turísticas, 29*, 57–80. https://doi.org/10.14198/INTURI.25478

BeCountry. (2025). *The most beautiful experiences at Italian farms*. https://becountry.it/en

Boucher, S., Flynn, A., & Lowe, P. (2017). The politics of rural enterprise: A British case study. In S. Whatmore, T. Marsden, & P. Lowe (Eds.), *Rural enterprise: Shifting perspectives on small-scale production* (pp. 120–140). Routledge.

Brouder, P. (2019). Creative tourism in creative outposts. In N. Duxbury & G. Richards (Eds.), *A research agenda for creative tourism* (pp. 57–68). Edward Elgar Publishing.

Capineri, C., & Romano, A. (2021). The platformization of tourism: From accommodation to experiences. *Digital Geography and Society, 2*, 100012.

Cardoso, M., Silva, G., & Fernandes, C. (2014). Traditional gastronomy events as tourist experiences: The case of Santarém gastronomy festival (Portugal). In G. Richards, L. Marques, & K. Mein (Eds.), *Event design* (pp. 181–197). Routledge.

Castells, M. (2000). *The rise of the network society* (2nd ed.). Blackwell Publishers.

Ciao Puglia Travel. (2025). Ciao Puglia Travel. https://www.ciaopugliatravel.com/

Clemetsen, M., Stokke, K. B., Barane, J., & Haraldseid, T. (2021). From tourist destination to local meeting place: Enhancing visitor experiences and social resilience in rural communities. In P. Fredman (Ed.), *Nordic perspectives on nature-based tourism: From place-based resources to value-added experiences* (pp. 50–63). Edward Elgar Publishing.

Cloke, P. L. (2007). Creativity and tourism in rural environments. In G. Richards & J. Wilson (Eds.), *Tourism, creativity and development* (pp. 59–69). Routledge.

Crocus Project. (2025). *Cultural and creative tourism in rural and remote areas*. Crocusproject.eu.

Divakaran, S. (2017). *Inside Airbnb's hyperlocal content strategy*. https://digitaluncovered.com/inside-airbnbs-hyper-local-content-strategy/

Droli, M. (2019). *The Albergo Diffuso model: Community-based hospitality for a sustained competitive advantage*. Walter de Gruyter.

European Commission. (2025). *Report on the long-term vision for rural areas*. https://rural-vision.europa.eu/index_en

Fernandes, C., & Richards, G. (2021). Developing gastronomic practices in the Minho region of Portugal. *Acta Geographica Slovenica, 61-1*, 141–152. https://doi.org/10.3986/AGS.9370

Fusté-Forné, F. (2019). Seasonality in food tourism: Wild foods in peripheral areas. *Tourism Geographies, 24*(4/5), 578–598.

Gyimóthy, S. (2017). The reinvention of terroir in Danish food place promotion. *European Planning Studies, 25*(7), 1–17.

Herslund, L. (2012). The rural creative class: Counterurbanisation and entrepreneurship in the Danish countryside. *Sociologia Ruralis, 52*(2), 235–255.

Hill, I., Manning, L., & Frost, R. (2021). Rural arts entrepreneurs' placemaking–how 'entrepreneurial placemaking' explains rural creative hub evolution during COVID-19 lockdown. *Local Economy, 36*(7–8), 627–649.

Horjan, G. (2011). Traditional crafts as a new attraction for cultural tourism. *International Journal of Intangible Heritage, 6*(1), 45–56.

Huhmarniemi, M., Kugapi, O., Miettinen, S., & Laivamaa, L. (2021). A sustainable future for creative tourism in Lapland. In N. Duxbury, S. Albino, & C. P. de Carvalho (Eds.), *Creative tourism: Activating cultural resources and engaging creative travellers* (pp. 239–253). CABI.

James, L., Halkier, H., Bradač, B., Ivanonv, S., Ivanova, M. G., Jurdana, D. S., Lind, A. L., Milfelner, B., Onderwater, L., Papp, Z., Raffay-Danyi, Á., Richards, G., Urdan, E., Saarinen, J., Spoljaric, T., & Zadel, Z. (2024). *CCT business models – CROCUS deliverable 2.2*. Aalborg University.

Khovanova-Rubicondo. (2012). Cultural routes as a source for new kind of tourism development: Evidence from the Council of Europe's programme. *Progress in Cultural Heritage Preservation – EUROMED, 2012*, 8–88.

Korez-Vide, R. (2017). Storytelling in sustainable tourism management: Challenges and opportunities for Slovenia. *Journal of Advanced Management Science, 5*(5), 380–386.

Leban, M., Errmann, A., Seo, Y., & Voyer, B. G. (2024). Mindful luxury: A case of The Faroe Islands. *Tourism Management, 104*, 104929.

Leer, J. (2016). The rise and fall of the new Nordic cuisine. *Journal of Aesthetics & Culture, 8*(1), 33494.

Leick, B., Gretzinger, S., & Roddvik, I. N. (2023). Creative entrepreneurs and embeddedness in non-urban places: A resource exchange and network embeddedness logic. *International Journal of Entrepreneurial Behavior & Research, 29*(5), 1133–1157.

Lonardi, S., Buffa, F., & Martini, U. (2024). Digital nomadism and tourism development: Stakeholders' perceptions of an inner area. *SINERGIE, 2024*(2), 91–107.

Long, P., & Lane, B. (2000). Rural tourism development. In D. W. Lime (Ed.), *Trends in outdoor recreation, leisure and tourism* (pp. 299–308). CABI.

Noivo, M. A., Lopes Dias, Á., & Jiménez-Caballero, J. L. (2022). Connecting the dots between battlefield tourism and creative tourism: The case of the Peninsular War in Portugal. *Journal of Heritage Tourism, 17*(6), 648–668.

Paniagua, Á. (2016). New rurality, events, and festivals in selected places of rural Spain. *Sociální studia/Social Studies, 2*(2016), 113–129.

Perkins, H. C. (2006). Commodification: Re-resourcing rural areas. In P. J. Cloke, P. Mooney, & T. Marsden (Eds.), *Handbook of rural studies* (pp. 243–257). Sage.

Perkins, H. C., Mackay, M., & Massacesi, C. (2023). Interpreting the 'rural mural' phenomenon: Creative place-making in rural towns in south and mid Canterbury, Aotearoa New Zealand. *Journal of Rural Studies, 103*, 103129.

Pine, B. J., & Gilmore, J. H. (1999). *The experience economy*. Harvard Business Press.

Richards, G. (2020). Designing creative places: The role of creative tourism. *Annals of Tourism Research*, 85. https://doi.org/10.1016/j.annals.2020.102922

Richards, G. (2021a). Evolving research perspectives on food and gastronomic experiences in tourism. *International Journal of Contemporary Hospitality Management, 33*(3), 1037–1058.

Richards, G. (2021b). Developing craft as a creative industry through tourism. *Brazilian Creative Industries Journal, 1*(1), 1–20.

Richards, G. (2024). The curatorial turn in tourism and hospitality. *International Journal of Contemporary Hospitality Management, 36*(13), 19–37. https://doi.org/10.1108/IJCHM-06-2023-0905

Richards, G., & Raymond, C. (2000). Creative tourism. *ATLAS News, 23*(8), 16–20.

Richards, G., James, L., & Halkier, H. (2024). *Concepts and methods for cultural and creative tourism in rural and remote areas. Crocus project deliverable D2.1*. Aalborg University.

Richards, G., Ashton Adie, B., Garibaldi, R., Halkier, H., Ivanov, S., Ivanova, M., James, L., Luna Lind, A., Milfelner, B., Mlakar, A., Onderwater, L., Palang, H., Palang, H., Pozzi, A., Raffay, A., Rudan, E., Saarinen, J., Šegota, T., Smolčić Jurdana, D., Špoljarić, T., Terziyska, I., Turnsek, M., Zadel, Z., & Zocchi, D. (2025). Cultural and creative tourism in rural and remote areas: European perspectives.*Tourism, 73(4)*. https://doi.org/10.37741/t.73.4.12

Ritzer, G. (1999). *Enchanting a disenchanted world: Continuity and change in the cathedrals of consumption*. Sage.

Schuhbauer, S. L., & Hausmann, A. (2022). Cooperation for the implementation of digital applications in rural cultural tourism marketing. *International Journal of Culture, Tourism and Hospitality Research, 16*(1), 106–120.

Shafernich, S. M. (1994). Open-air museums in Denmark and Sweden: A critical review. *Museum management and Curatorship, 13*(1), 9–37.

Śledziewska, K., & Włoch, R. (2021). *The economics of digital transformation: The disruption of markets, production, consumption, and work*. Routledge.

Stiman, M. (2020). Second homes in the city and the country: A reappraisal of vacation homes in the twenty-first century. *International Journal of Housing Policy, 20*(1), 53–74.

Tamborrino, R. R. M. (2020). Surveying and actioning the tangible and intangible dimensions of rural heritage in RURITAGE H2020 project. In *World heritage and contamination* (pp. 616–623). Gangemi Editore International Publishing.

Törnberg, P. (2022). Platform placemaking and the digital urban culture of Airbnbification. *Urban Transformations, 4*(1), 3.

Trono, A., & Castronuovo, V. (2021). The via Francigena del Sud: The value of pilgrimage routes in the development of inland areas. The state of the art of two emblematic cases. *Revista Galega de Economía, 30*(3), 1–18.

Chapter 11
Adaptive Reuse of Historic Urban Spaces: Challenges in Interior Redesign for Tourism

Maria Tămăşan

Abstract The adaptive reuse of historic urban spaces for tourism brings both opportunities and challenges in interior design, requiring a careful balance between heritage conservation, contemporary functionality, and aesthetics. As cities seek to preserve their cultural identity while accommodating modern visitors, designers must address complex functional, material, spatial, and regulatory demands. The added value of this research lies in highlighting interior design's role as a mediator between past and present, contributing to urban development and cultural continuity.

The study investigates how contextual analysis shapes interior design strategies for transforming historic buildings into tourism-oriented spaces. Through a selective and comparative analysis of adaptive reuse projects in Romania, the research identifies best practices that maintain authenticity while catering to visitors' needs. The methodological approach combines case study analysis, on-site observation, critical interpretation, and literature review, focusing on the ratio between the degree of architectural intervention in a given space and the cultural programming that caters to an attractive user experience.

Key findings show that successful interventions are grounded in a deep understanding of the historical context and placemaking, as well as in a phenomenological grasp of what constitutes tourist attractions. Designers must preserve original architectural features while introducing new functions in ways that are respectful, reversible, and clearly distinguishable from the historical fabric.

The main conclusion emphasizes that adaptive reuse, when guided by contextual insight and design sensitivity, enhances the cultural, educational, and economic value of historic sites. Thus, they become assets for tourism and hospitality developments and a context for events that attract the public. Interior design becomes not only a tool for aesthetic transformation but also a means of placemaking, stability, and historical storytelling. This approach fosters emotional connections between visitors and spaces, reinforcing local identity and increasing the uniqueness and attractiveness of repurposed urban heritage.

M. Tămăşan (✉)
Faculty of Design, 'Aurel Vlaicu' University of Arad, Arad, Romania
e-mail: maria.tamasan@uav.ro

A. L. Negrusa, M. M. Coroş (eds.), *Innovation, Sustainability, and Growth in a VUCA Environment*, Springer Proceedings in Business and Economics,
https://doi.org/10.1007/978-3-032-11639-0_11

Keywords Historic urban site · Adaptive reuse · Valorization · Interior redesign · Placemaking

11.1 Introduction

This chapter investigates the impact of how interior designers balance adaptive reuse of historic urban buildings intended for tourism by maintaining contextual fidelity with innovation and cultural programming, in order to emphasize the inherent character of the space. Adaptive reuse of historic spaces has become a key strategy for sustainable development in urban areas. Historic buildings lie at the heart of both *social memory* and historic urban centers, where much of the population is concentrated (Takva et al., 2023). Over time, changes in lifestyles and needs have exposed many of these structures to the risk of extinction—either through physical deterioration or, more commonly, through a form of non-physical extinction caused by aesthetic uniformization. This homogenization leads to a gradual erosion of their identity within collective memory and, without strong marketing campaigns, their attractiveness for future investment diminishes. Consequently, professionals must advocate for greater awareness and responsibility in interventions on historic buildings, particularly within the insufficiently regulated field of interior design.[1] In the context of interior design for historic buildings, the creative process entails a continuous negotiation between the memory of place, the identity embedded within historical spaces, and the functional and aesthetic requirements of the present. The interior designer operates at the intersection of art and science, perception, and technology, acting as an interpreter of the subtle signals conveyed by the existing environment. Moving between past and future, between the intuition of origins and the drive for innovation, the designer transforms interiors into layered cultural and emotional palimpsests—stages for both visible and invisible narratives.

Within this framework, the adaptive reuse of historic interiors for tourism purposes acquires new layers of meaning. Such projects require designers to preserve and reveal historical value while integrating contemporary functions and technologies. Interior spaces are not passive backdrops, but active agents shaping the atmosphere and experience of a site. Through contextual analysis, material sensitivity, and narrative-driven strategies, interior designers can craft spaces that mediate between memory and transformation, reinforcing the authenticity and distinctiveness of place.

Each phase of transformation in historic urban centers inevitably involves a series of conflicts. Every renewal implies a degree of erasure—whether partial or complete—yet each historical layer embodies specific values, ideologies, needs, and social statuses. Through research and critical interpretation, these hidden layers can be unfolded into narratives unique to each site, made accessible to the public

[1] In comparison with architectural practice in Romania.

through careful and authentic design interventions. By acknowledging the layering of time and space, interior design projects construct a temporal–spatial continuum that invites both visual and cognitive engagement with history. Every place possesses a unique historical trajectory; thus, the true, irreproducible character of a space can emerge only through this attentive approach. In this way, spaces are transformed from neutral coordinates into meaningful places, acquiring significance that can be strategically leveraged in the fields of hospitality and tourism. Design becomes a fundamental component of branding and marketing strategies, enhancing the identity, appeal, and competitiveness of the site.

In Romania, interior design is a relatively recent professional field; only in recent years has there been a broader public willingness and financial capacity to engage specialized services for interior design projects. Within this framework, interior design interventions in historic spaces represent both a sensitive undertaking and an underexplored niche, as they require interdisciplinary expertise encompassing restoration practices, material and finish conservation, technical installations, historical research, and the critical recognition of valuable elements requiring preservation. These factors are both fundamental and restrictive to the artistic proposal: while some practitioners may perceive them as limitations, others embrace them as creative challenges, prompting a deeper pursuit of the *poesis* of space. This can only be truly achieved by unfolding the layers—both physical and symbolical—that define the space in a thorough process similar to archaeology (Istiani, 2022).

In the context of an increasingly dynamic tourism and hospitality market, marked by intense competition for clients and visitors, investors are constantly seeking innovative and alternative concepts for the adaptive reuse of spaces within city centers. However, the misinterpretation of trends, combined with a superficial understanding of the historic value embedded in certain areas, buildings, or spaces, has often led to the proliferation of repetitive interior designs. Although these interiors may differ in materials, lighting, or thematic choices, they frequently lack a genuine connection to the specific site, resulting in designs that fail to evoke emotional engagement from visitors.

From an architectural and cultural perspective, the challenge lies in finding a balance between design innovation, historical integrity, and public engagement.

This chapter explores means of achieving that balance by analyzing three case studies in Brașov and Arad, Romania, that reflect different approaches to adaptive reuse: one involves a significant architectural intervention guided by heritage-conscious design; the second is a commercially driven project with minimal preservation; and the third case presents a historical structure nearly untouched, yet revitalized through recurring public events and a consistent cultural concept anchored in the history of the place. These cases offer a nuanced view of how adaptive reuse can operate not just through physical transformation but also through curation, activation, and programming.

This chapter addresses the following research question:

How can the balance between physical design intervention and cultural programming guide successful adaptive reuse of historic urban spaces?

11.2 Literature Review

The adaptive reuse of historic urban spaces for tourism purposes involves a judicious balance between heritage preservation, integration of modern facilities and functions, aesthetics, and memorable user experience. The latter is usually achieved by *cultural programming*, which aims to attract the public and bring those spaces to its attention, thus enhancing their cultural and economic viability. It is safe to say that adaptive reuse has evolved from pragmatic conservation strategies to a complex holistic design and management process that takes into account and correlates key concepts such as context, user profile and expectations, historic value, preservation and musealization practices, modern lifestyle, visitor-oriented experiences, place-making, and narratives.

In *Building a New Heritage: Tourism, Culture and Identity in the New Europe*, Ashworth and Larkham (1994) highlight the fundamental relationship between monuments and their surroundings, arguing that historic landmarks actively shape a place's identity (Ashworth & Larkham, 1994). This view further reinforces the idea that the built environment is not a passive container, but an active agent in constructing meaning, memory, and social narratives, which is critical for tourism contexts where site-specific experiences drive visitor engagement. Therefore, the authors suggest that besides preserving the historic structures and modifying to adapt them to recent needs, the identity-forming elements must be saved and put to good use in order to maintain the cultural significance of the place. Cultural programming amplifies this by creating dynamic, visitor-oriented activities.

According to Macdonald (2013), Joachim Ritter and Hermann Lübbe coined the concept of "musealization" between the '60s and the '80s, noting the fast transition of the present into the past, due to accelerated urban and technological development. By extrapolating his point of view on museum practices, one can argue that historic spaces should function as reinterpreted museums, offering didactic, aesthetic, and affective experiences. Visiting such spaces becomes a multi-layered experience: an aesthetic one, offered by the quality of the interior design; a didactic one, by acquiring knowledge about the chosen context and its past; and an affective one, achieved through the internalization of intimate details that ensure the uniqueness of the experience and increase the likelihood of the visitor's return. These experiences can be achieved not only by design-driven interventions on the surfaces but also by cultural programming, which curates events and activities that foster emotional connections with the space and increase the likelihood of a return.

Plevoets and Van Cleempoel (2011, 2019) emphasize the importance of interior design as a distinct discipline within adaptive reuse, one that mediates between heritage conservation and contemporary lifestyles. They argue that interiors are active carriers of meaning, atmosphere, and experiential depth. Their work underlines the importance of design strategies that respect historical layers, even though they introduce new, legible, reversible, and relevant contemporary elements, especially for tourist or public function, where expectations demand both authenticity and functionality. In a more recent study, Augustiniok, Houbart, Plevoets, and Van Cleempoel

(2023) provide a comprehensive framework for adaptive reuse by examining the interplay between heritage values (e.g., architectural, historical, and cultural) and contemporary design values (e.g., spatial, functional, and experiential). They present a qualitative study of two projects in Belgium—Predikherenklooster in Mechelen and C-Mine in Genk, emphasizing how design strategies leverage the existing spatial qualities, namely, the attic in Predikherenklooster and the turbine hall in C-Mine in Genk, in order to create meaningful visitor-oriented experiences. The study also acknowledges the importance of historic value recognition, hierarchization, and interpretation in the process of design-driven decisions that aim at an equilibrium between usability, historic material preservation, space symbolistic, and enhanced user experience. The authors emphasize that a detailed understanding of heritage values ensures interventions that enhance the cultural value of the space.

In a similar line of interpretation, although not focused on historic buildings, Andron (2024) introduces the term "surface semiotics," suggesting that elements such as colors, layers of paint, materials, and traces of the past contain valuable information about the context that determined them. The surface semiotics, as interpreted by various scholars, as well as Pleovets and Van Cleempoel, is an important starting point for a well-structured and authentic narrative. In tourism contexts, cultural programming can build on these surfaces by curating events that interpret and amplify their narratives, enhancing visitor immersion. For example, in the Humanitas Bookstore in Braşov, one of the proposed case studies in this chapter, restored original frescoes could be easily complemented by cultural events like book readings or guided tours, which deepen the visitor experience. To further explore the role of cultural programming in space reactivation, Markusen and Gadwa (2010) argue that cultural programming is a critical strategy for revitalizing historic urban spaces. They highlight how arts-based events, such as festivals, performances, or temporary exhibitions, can transform underutilized spaces into vibrant cultural hubs, attracting tourists and fostering community engagement. Their analysis of creative placemaking emphasizes that programming should be context-specific, drawing on the site's historical and cultural narratives to create authentic experiences that resonate with diverse audiences. And that can be only achieved by discerning the valuable historic layers, preserving and revealing them to the public by restoration or interpretative integration.

Similarly, Richards and Wilson (2006) explore cultural tourism and the role of creative approaches in reactivating urban and rural spaces, including historic ones, as well as the importance of the narrative to give coherence to the fragmented pieces of information one can acquire through a cultural experience. They argue that *creative* tourism is a branch of cultural tourism with high potential to create unique memorable place-linked memories for the visitor. Thus, cultural programming should focus prevalently on creative activities (*creative tourism)* where visitors are given the opportunity to develop their own creative potential by engaging in various activities, as well as on such activities that are linked to the location, so that the learning experience (*educational tourism*) is directly linked to the visited location.

Brooker and Stone (2004, 2018) further highlight the designer's interpretative role in creating new spatial narratives that respect the historical layers of a site while introducing modern interventions that are readable, reversible, and relevant. They advocate for interventions that are distinguishable from the original fabric, ensuring historical integrity. This theoretical perspective aligns with international charters on heritage conservation (such as the Venice Charter and the Burra Charter), which emphasize the importance of minimal, respectful intervention, ensuring legibility and integrity of historic sites. Adaptive reuse in the context of tourism adds another layer of complexity, as it must respond not only to spatial and regulatory demands but also to the experiential and emotional expectations of visitors.

Collectively, these sources underline the complexity of adaptive reuse in historic urban spaces for tourism, where interior design and cultural programming work synergistically to preserve authenticity while meeting contemporary demands and the need for constant innovation. Interior design provides the physical framework for storytelling and placemaking, while cultural programming activates spaces through curated experiences that engage visitors emotionally and intellectually. This interdisciplinary approach, integrating historical research, restoration expertise, innovative design, and strategic programming, ensures that repurposed spaces remain culturally and economically viable in competitive tourism markets.

While existing research offers robust theoretical models for conservation, transformation, and tourism integration, few studies explore adaptive reuse through the lens of a practicing architect evaluating reuse strategies based on spatial design and cultural activation. Moreover, regional studies from Central and Eastern Europe remain underrepresented. This chapter contributes to filling that gap by analyzing three contrasting cases from Braşov and Arad, Romania, and proposing a design framework based on the degree of physical intervention and the extent of cultural programming.

11.3 Research Methodology

This research adopts a qualitative, design-based case study approach to explore how architectural intervention and cultural programming influence the reuse of historic urban spaces for tourism and public engagement. Two case studies from the historic center of Braşov, Romania, and another from Arad, were selected based on the following criteria:

- Recent implementation of adaptive reuse interventions (within the last two decades)
- Clear transformation of function
- Physical accessibility and direct observation by the author
- Distinctive design strategies reflecting different interpretations of the historic context

The three distinct approaches to adaptive reuse represented by the selected examples have the following particularities: one is featuring significant architectural and restoration interventions with great care for correctly restoring the original historic decoration, integrating it in the contemporary design, another involving a functional private redevelopment with limited heritage retention, and a third where the historic structure remains physically untouched but is reactivated through a vibrant cultural program.

Data were collected through direct field observation, on-site documentation, review of architectural drawings (where available), and public discourse analysis from architects and stakeholders. To guide the comparative analysis, the study proposes a conceptual framework based on the following criteria:

- The degree of physical intervention in the built fabric
- The extent of cultural programming as a tool for reactivation and public use
- The architectural coherence of the interior space.[2]
- The historical readability
- The observed visitor engagement
- Urban integration
- Narrative coherence—alignment between concept, history, and design language
- Uniqueness of concept—how creative or how repeatable is the main concept of the design

Each case is positioned in a comparative matrix table (see Table 11.1), illustrating how different combinations of design strategies can achieve varying outcomes in heritage valorization. This methodological tool serves not only as a comparative device but also as a potential guide for future design decisions, allowing practitioners to evaluate reuse scenarios beyond conventional typologies.

Site observations were carried out in situ (Brașov, 2018, 2022; Arad, 2012–present), with attention to materiality, lighting, circulation, interpretive features, and spatial layering on one hand and the type of activities and the public on the other. These were supplemented by literature, photographs, and available project documentation. The objective is not to generalize but to extract transferable principles for interior design in heritage tourism contexts.

Case Study 1: Humanitas Bookstore—Brașov

The first case study is the Humanitas Bookstore at no. 16, Piața Sfatului, Brașov, a complex intervention in a historic ensemble originally part of the "Apfelmarkt" (Apple Market) area, visited in 2022. The restoration project began in 2004 and involved the consolidation, restoration, and adaptive reuse of the historic building, now integrated into the cultural life of the city. Originally, a patrician residence later converted into a multi-family dwelling and commercial space. The inauguration of the Humanitas Bookstore marks the symbolic moment when the building re-enters

[2] As an architect specializing in heritage and design, the author also brings professional judgment into the qualitative evaluation of spatial strategies and heritage coherence.

Table 11.1 Comparative matrix table (design dimensions)

Criteria	Case 1: Humanitas Bookstore Braşov	Case 2: Dr. Jekelius Pharmacy Cafe Braşov	Case 3: Joy's Literary Caffe
Degree of physical intervention	High—preservation	Medium—new	Low
Cultural programming	Medium	Low	High
Architectural coherence	Strong	Fragmented	Preserved
Historical readability	Enhanced	no historic authenticity	Intact
Visitor engagement (observed)	Active and targeted	Active and diverse	Informal, diverse, and fluid
Urban integration	Anchored in community	Anchored in community	Anchored in community
Narrative coherence	High	Medium/low	Medium/high
Innovative concept	Medium	High	Low

the public sphere after decades of neglect. The interior design carefully preserves and highlights the building's historic features—such as mural paintings, sculpted stonework, and original plaster—while introducing new functions with minimal visual disruption. All furniture, lighting, and accessories were custom-designed to respect the spatial and material identity of each room, allowing the historic character of the space to remain dominant and legible within its new use (Bertleff et al., 2014).

All contemporary interventions are carefully distinguished from the historic structure: for instance, the new parquet flooring is separated from the walls by a strip of decorative gravel—a detail that is both aesthetic and functional, ensuring continuous ventilation of the original structure, which remains vulnerable to moisture rising from below. Another striking feature for visitors is the extensive presence of restored frescoes, revealed and preserved across the walls and vaults, even at the apparent risk of competing with the displayed merchandise for the visitor's attention. However, rather than presenting a conflict, this coexistence enhances the overall experience, transforming the act of shopping into a layered cultural encounter.

Cultural programming in this case is focused on book promotion and commerce, acting as a marketing strategy for the library, not so much as a tourist attractor. However, the unique character and atmosphere of the space prevail and make the visit memorable (Fig. 11.1).

Case Study 2: Dr. Jekelius Pharmacy Café—Braşov

The second case study focuses on the former Dr. Jekelius Pharmacy Café in Braşov, now Viva la Vida bistro hostel, visited in 2018 before its closure and rebranding in 2021. Although situated within the same historic urban area as the Humanitas

Bookstore, the interior design and branding approach of the café is distinctly different, yet equally effective—if not more so. The café's theme draws inspiration from the building's previous function as a pharmacy, specifically "La Speranța," which was operated by the Jekelius family. Dr. Ferdinand Jekelius, a notable figure in the late nineteenth century, played a central role in its legacy. The building itself dates back to the eighteenth century and showcases architectural elements characteristic of that period. Unfortunately, much of the original furniture was lost due to expropriation during the communist era and other subsequent events following the regime change (Anon., fără an). Despite this, every detail—from cabinets to chairs,

Fig. 11.1 Interiors in Humanitas Bookstore in Brașov. (Courtesy of Viorel Dragomir, 2025)

finishes, and objects—references the pharmacy, creating a surreal experience for visitors. Though not original, the setting transports customers between past and present, offering a unique ambiance that preserves the place's identity and honors a significant figure in the city's history.

What is remarkable in this case is the fact that the concept itself, even though not 100% authentic, acted as an attractor for the public, making any other form of marketing unnecessary. The uniqueness of having the opportunity to use pharmacy-specific instruments as glasses for drinks and to entertain the illusion of a hybrid space—not a vintage pharmacy, nor a bar—is sufficient to attract the public. The space is still remembered in the first form, despite the rebranding in 2021, which is incontestable proof of the importance of a well-documented, locally anchored design concept (Fig. 11.2).

Case Study 3: Joy's Literary Caffe—Arad

This case study explores Joy's Literary Caffe (Cafeneaua Literară Joy's) in Arad, Romania, as an exemplar of an alternative approach to adaptive reuse. The coffee shop is housed within the former residence of Jacob Hirschl, recognized as one of Arad's earliest residential structures. Adjacent to this building, Hirschl constructed the old city theater between 1816 and 1817, notable as Romania's first permanent theater, which remained operational until 1873. Presently, the theater stands largely abandoned, though it sporadically serves as a venue for alternative cultural events. The coffee shop itself has undergone minimal interior refurbishment, with its primary focus directed toward the courtyard, which emerges as the establishment's principal draw (Fig. 11.3).

The courtyard, an exterior space enveloped by surrounding buildings, offers year-round usability. It accommodates approximately 50 individuals on the ground floor, with additional seating available on terraces that connect to the first floor.

Fig. 11.2 Interiors in the former Dr. Jekelius Pharmacy Café in Braşov, now Viva la Vida bistro hostel. (Source Orsi Bohemian—foto&video: https://www.facebook.com/bohemianstudio/photos/t.100063671525745/2497319893639636/?type=3)

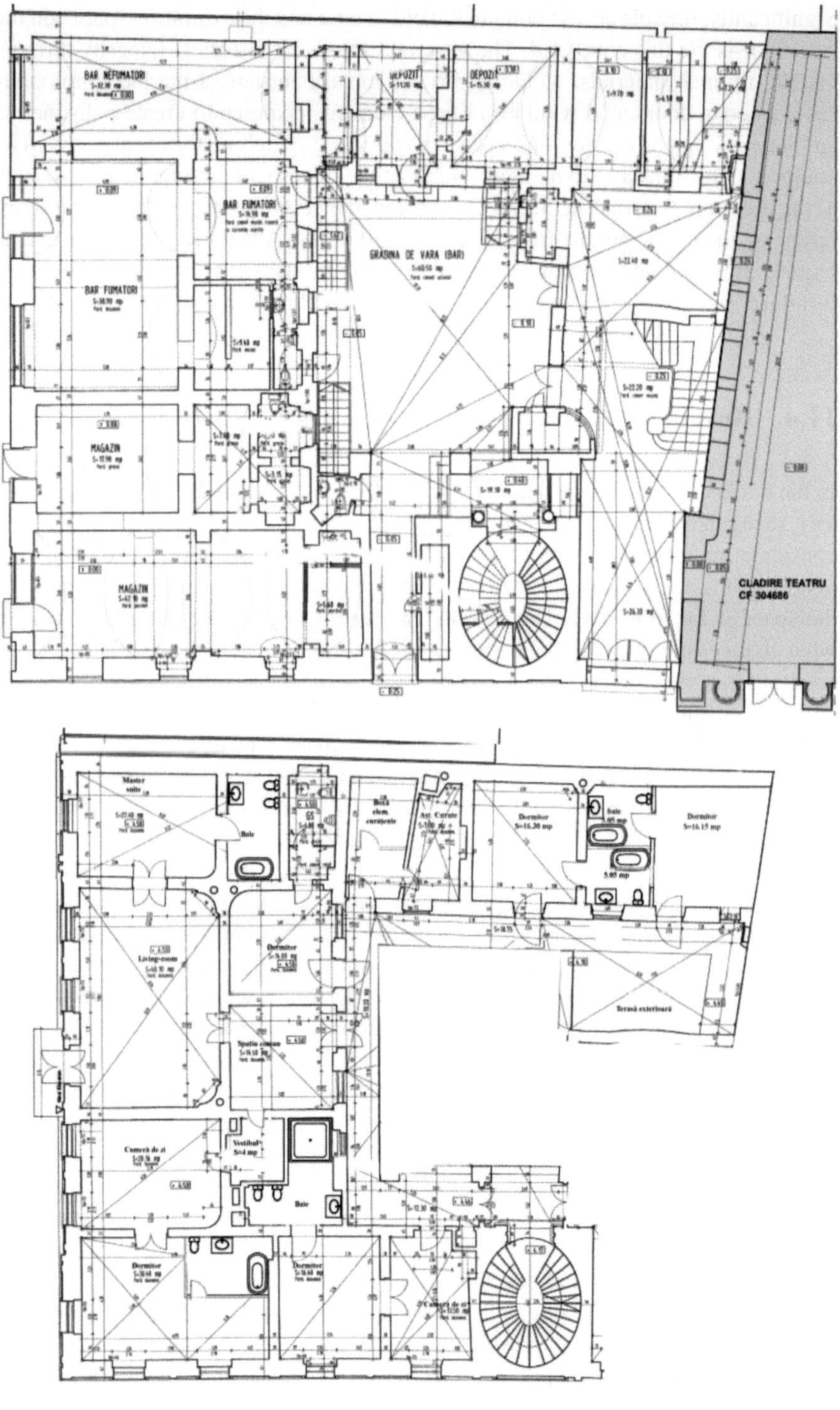

Fig. 11.3 Floor plans of the Hirschl residence (above—Ground floor. Below—first floor). (Source: s.c. Tara Plan s.r.l.)

Significantly, the sole access point to the old theater and the exhibition spaces on the first floor is via this courtyard. The space preserves its historical integrity through visible original materials, including bricks, metal banisters, a marble spiral staircase, wooden doors, and windows. These elements coalesce to create a distinctive ambiance, frequently prompting visitors to ponder the site's historical narrative. The courtyard functions as a dynamic venue hosting an array of cultural activities, such as the literature festival The Secret Discussion, dance and music performances, conferences, theatrical productions, and art exhibitions. Adaptation for these events requires minimal intervention, achieved through the strategic placement of posters, furniture, projection screens, and photographs that chronicle past activities.

11.4 Results and Discussions

In the bookstore, the design approach acknowledges the layering of the space's history, recognizing that the new function adds another layer to the existing ones. This consideration makes the perceptive and attentive visitor a participant in the ongoing history of the space. This idea can be seen as a subtle promise that the evolution of the space is far from over. The curious and receptive observer will likely wonder what changes are yet to come, fostering a long-term connection with the space and ensuring that visitors return.

In the chapter "Surface Semiotics" from *Urban Surfaces, Graffiti, and the Right to the City*, Sabina Andron explores the notion that "discourses in cities are also discourses about cities, about values, needs, and aspirations." According to Andron, reading the city involves not only observing but also understanding the connections between signs, inscriptions, and surfaces. She further argues that "surfaces are thick and alive" (Andron, 2024), acting as carriers of meaning and information. By analogy, every surface—whether interior or exterior—can be interpreted in this way. Consequently, the principles of interpreting urban surfaces at a macro level can also be applied to the scale of a single building or even a specific interior space. However, it is crucial to emphasize that the information derived from reading and analyzing surfaces and forms must be anchored in a holistic understanding of the broader context to maintain its relevance and significance. Colors, materials, shapes, creases from past installations, even carved inscriptions, discoloration from light, and darkened areas from furniture or decoration—all of these can serve as valuable tools in constructing a comprehensive interior design concept. This is why, in the case of the bookstore, the design process took longer but is more likely to have lasting value.

At the Dr. Jekelius Pharmacy Café, the visitor is transported into a neutral space that exists between the present and the past. In contrast, the concept in the Pharmacy Café was more explicit and accessible to a broader public. The approach in the bookstore, however, is subtler and appeals primarily to those with a background in restoration practices or an interest in heritage preservation and interpretation. To an untrained eye, the exposed historic material might appear as an eccentric form of wall decoration. Nevertheless, both approaches are valid and rooted in their respective contexts. The key difference lies in the fact that the interior design of the bookstore follows a more scholarly approach, with decisions based on solid, interdisciplinary research. This ensures the durability of the design and its impact. Even if the function of the space were to change in the future, the uncovered historic layers on the walls have a strong likelihood of persisting. Coincidentally, or perhaps subliminally, the concept of "reading" the building—an idea introduced in the earlier discussion on the semiotics of surfaces—occurs naturally within the Humanitas Bookstore. The design emphasizes layers of meaning, which invite visitors to engage with the space in a multi-dimensional way. Each surface, material, and historical element tells a story, guiding the visitor through an ongoing process of interpretation. This act of "reading" the building creates a dynamic relationship between the past and the present, one in which the observer is not a passive participant but an active agent in the construction of meaning. Whether consciously or not, the visitor becomes part of the evolving narrative of the space, making the building's history an integral part of their own experience.

A distinct approach can be observed in the case of Joy's Literary Caffe, where the connection between the space and the visitor relies on the awareness and curiosity of the latter. However, subliminally, those who partake in the activities and the events happening there are prone to remember and return.

After in situ observation and data correlation, the results of the study have been synthesized in the matrix in Fig. 11.4, offering an overview of different scenarios of successful space reuse achieved. One outcome of this analysis is that there is no recipe for success, but there is a fine balance between the originality of the concept, the degree of physical design-driven intervention on the space (which ensures historical readability and the narrative coherence), and cultural programming, which can generate natural or determined visitor engagement.

11.5 Conclusions

This study highlights how the adaptive reuse of historic interiors can support cultural tourism, identity formation, and community activation—but only when grounded in contextual awareness and interpretive depth. The comparative analysis reveals that design intervention alone is insufficient: even minimal architectural change can foster strong public engagement when paired with coherent cultural programming. Conversely, a richly restored site may fail to engage without meaningful activation.

The matrix model developed here shows that success in adaptive reuse hinges on three interrelated dimensions: the degree of physical intervention, the extent of cultural programming, and the narrative coherence of the concept. Case 1 exemplifies an academically rigorous design strategy rooted in historical restoration. Case 2, with limited original fabric, demonstrates how thematic storytelling and sensory design can still generate public interest. Case 3 illustrates the potential of minimal intervention amplified by high-quality cultural programming.

Fig. 11.4 Courtyard of the Hirschl residence. (Source: author's personal archive)

For designers working with historic urban sites, this study suggests the need for an **integrated strategy**: architecture and programming must work hand in hand. Furthermore, this research suggests that the adaptive reuse of historic buildings requires an interdisciplinary approach, integrating expertise from restoration practices, historical research, and contemporary design techniques. Such collaboration ensures that interventions remain respectful of the building's legacy while adapting it for new functions. In this sense, adaptive reuse provides a resilient framework capable of responding to the volatility, uncertainty, complexity, and ambiguity (VUCA) that characterize contemporary urban environments, allowing spaces to evolve with changing needs while preserving their core identity. Moreover, visitor engagement is not simply a byproduct of design—it is an indicator of its success. Adaptive reuse, in this view, is not preservation in disguise, but a forward-looking cultural practice that reactivates the past through meaningful design. This can be summed up in three key findings that suggest the following:

- Authenticity in adaptive reuse may be material (preserving original features) or conceptual (preserving narrative essence).
- Design decisions must be guided by the intended user experience and the site's heritage potential.
- Emotional engagement and spatial memory are core components of successful reuse in tourism contexts.

Ultimately, interior design acts as a mediator between the past and present, providing tools not only for visual transformation but also for cultural translation and economic regeneration. When informed by interdisciplinary insight and contextual depth, adaptive reuse can yield spaces that are both meaningful and marketable—fostering continuity without stagnation and innovation without erasure.

References

Andron, S. (2024). *Urban surfaces, graffiti, and the right to the city* (pp. 9–10). Routledge.
Ashworth, G. J., & Larkham, P. (1994). *Biulding a new heritage. Tourism, culture*. Routledge.
Augustiniok, N., Houbart, C., Plevoets, B., & Van Cleempoel, K. (2023). Adaptive reuse of built heritage: Conserving and designing with values. *Journal of Cultural Heritage Management and Sustainable Development, 15*(1), 24–41.
Bertleff, J., Oprea, D. & Creosteanu, L., 2014. Librăria „Humanitas", Braşov (Casele Albrich-Hiemesch). s.l., s.n.
Brooker, G., & Stone, S. (2004). *Re-readings: Interior architecture and the design principles of remodelling existing buildings*. RIBA Enterprises.
Brooker, G., & Stone, S. (2018). *Context: The design work of interior architecture*. Fairchild Books.
Istiani, N. F. F. (2022). An archaeology of the interior. Case studies: The archaeology of obsolete stairs. *Indonesian Journal of Visual Culture Design and Cinema*.
Macdonald, S. (2013). *Memorylands. Heritage and identity in Europe Today*. Routledge.
Markusen, A., & Gadwa, A. (2010). *Creative Placemaking*. Available on-line at https://obserwatorium.miasta.pl/wp-content/uploads/2017/02/CreativePlacemaking_Ann-Markusen-Anne-Gadwa.pdf
Plevoets, B., & Van Cleempoel, K. (2011). *Adaptive reuse as a strategy towards conservation of cultural heritage: A literature review*. WIT Press.
Plevoets, B., & Van Cleempoel, K. (2019). *Adaptive reuse of the built heritage: Concepts and cases of an emerging discipline*. Routledge.
Richards, G., & Wilson, J. (2006). Developing creativity in tourist experiences: A solution to the serial reproduction of culture? *Tourism Management Issue, 27*(6), 1209–1223.
Takva, Y., Takva, Ç., Gökşen, F., & İlerisoy, Z. Y. (2023). Examination of contemporary additions made with adaptive reuse. *Online Journal of Art and Design*.

Chapter 12
Hospitality in Romanian Hospitals: A Modern Customer Experience

Ionuț Vida-Simiti

Abstract There is an emerging international concern about the concept of hospital hospitality as an important role in improving patient healthcare experience. So, this notion of healthcare hospitality that merges two service industries, the healthcare industry and the hospitality industry, began to attract the attention of scholars and practitioners. In Romania, based on Article 166 para. (1) of Law No. 95/2006, a hospital "is a health facility with beds," which means that hotel services are part of the essence of the hospital as a health unit. So, the Romanian hospitals can provide three kinds of hotel services: (a) basic hotel services (for which the hospital must be accredited by the National Agency for Health Quality Management), (b) superior hotel services, and (c) high comfort hotel services, respectively. These services may be provided to both patients and third parties. However, the beneficiaries of the hotel services provided by the hospital must contribute to the welfare of the patients hospitalized in that hospital, which is why Article 233 of Law No. 95/2006 expressly refers to the mothers of underage patients and, respectively, the accompanying persons of severely disabled persons. But, since the hotel service is an auxiliary service in hospitals, the juridical literature did not take into consideration this issue. In consequence, the aim of this study is to analyze these legal provisions in the framework of medical law, to identify situations of liability when the provision of hospitality services is not carried out properly, and to provide solutions not only to avoid them but also to improve customer experience. So, the criteria and requirements relating to hotel conditions for the assessment of hospitals are mentioned in Annex 1 of *the Standards for the Accreditation of Hospitals* of Order 446/2017 *on the approval of the Standards, Procedure and Methodology for the Assessment and Accreditation of Hospitals* and the way in which all these conditions must be met is described in the Accreditation Standards Manual for the Accreditation of Bed

---

I. Vida-Simiti (✉)
University of Medicine and Pharmacy "Iuliu Hațieganu", Cluj-Napoca, Romania
e-mail: ionut.vida-simiti@umfcluj.ro

A. L. Negrusa, M. M. Coroș (eds.), *Innovation, Sustainability, and Growth in a VUCA Environment*, Springer Proceedings in Business and Economics,
https://doi.org/10.1007/978-3-032-11639-0_12

Health Facilities of the National Agency for Quality Management in Health Care. But, the evaluation of hotel service delivery and especially the impact on the patient is very difficult to realize because, on the one hand, the Order 1501/2016 *on the approval of the implementation of the patient feedback mechanism in public hospitals* includes in Annex 1 a form in which discharged patients are invited to answer questions in order to improve health services, none of which refers to the accommodation conditions, and, on the other hand, the questionnaires in Annex 65 and 66 of the *Order of the President of the National Agency for Quality Management in Health Care No. 8/18.01.2018* include only very general queries on some hotel issues such as the existence of a locker or wardrobe where to store the clothes and personal belongings with which he/she came, or the provision of a set of pajamas, robes, slippers, matching, and bed linen from the hospital. However, for these hotel services, Article 166 para. (2) of Law No. 95/2006 establishes the hospital's responsibility, but without mentioning a special legal liability. That is why, in the case of military hospitals, the Ministry of Defense has tried to transfer through Article 42 para. (3) of the Annex to Order No. M 110/2009, the responsibility for the accommodation and food conditions from the hospital to the medical staff as a new form a malpractice liability.

Keywords Hospitality · Hospital · Medical law · Malpractice

12.1 Introduction

The notion of healthcare hospitality merges two service industries: the healthcare industry (mainly hospitals) and the hospitality industry (mainly hotels), because it combines complex service features of the two, on the one hand, medical treatment, qualification of doctors and paramedical staff to render quality healthcare services, accreditation of hospitals, and alternative health treatments provided by the healthcare industry and, on the other hand, comfortable amenities, a harmonious mixture of food, beverage, and shelter through restaurant and room service, nourishment, protocol, and easy access parking lot characteristics for the hospitality industry. That is why it has an important role in improving the healthcare experience, and it has begun to attract the attention of scholars and practitioners in different frontiers of healthcare, hospitality, health tourism, wellness tourism, medical tourism, and consumer behavior (Majeed & Kim, 2023).

12.2 Literature Review

In Romania, unlike in France, where the doctrine distinguishes between the healthcare contract and the hotel contract by which distinct main services are provided (Dorsner-Dolivet, 2006), the law regulates hotel service as only an auxiliary service to the main obligation to provide healthcare (Vida-Simiti, 2013). Since the hotel service is an auxiliary service, the law does not insist on the concrete manner in which it is provided. Maybe that is way the juridical literature did not take in consideration this issue with the exception of a particular case when it signaled an attempt is made to exempt the hospital from the hotel obligations of the civil code by shifting the responsibility to the doctor as in the case of military hospitals where the responsibility for accommodation and food conditions are placed on the commander, the head of the ward, and the attending physician (Vida-Simiti, 2018).

12.3 Research Methodology

According to Article 163 paragraph (1) of Law No. 95/2006 *on healthcare reform*, the hospital is defined as a health facility with beds, which, based on Article 166, provides, in addition to medical care, accommodation and food for patients, and under Article 193 para. (7) letter g. may also provide hotel services to third parties. The aim of this study is to analyze these legal provisions in the framework of medical law, in particular, and the Romanian Civil Code, in general, in order to identify situations of liability when the provision of hospitality services is not carried out properly and to provide solutions not only to avoid them but also to improve customer experience.

12.4 Results and Discussions

Based on Annex 1. Point 28 of Commission Regulation (EU) 2021/1901 of October 29, *2021 implementing Commission Regulation (EC) No. 1338/*2008 *of the European Parliament and of the Council of December 16, 2008 concerning statistics on expenditure and sources of financing in the field of health*, hospitals are defined as "approved establishments whose activity consists mainly in the provision of medical, diagnostic and treatment services, which include services provided by doctors, nursing care and other health services provided to inpatients, and specialized accommodation services required for inpatients, and which may also provide day care, outpatient and home health services." In national law, Article 166 para. (1) of Law No 95/2006 stipulates that a hospital "is a health facility with beds, of public utility, with legal personality, which provides medical services." It follows that hotel services are part of the essence of the hospital as a health unit.

Thus, according to Article 1 of Annex 2 of the *Procedure and Methodology for the Assessment and Accreditation of Hospitals* of Order 446/2017 *on the approval of the Standards, Procedure, and Methodology for the Assessment and Accreditation of Hospitals*, the assessment of hospitals covers several areas referred to as reference, each of which groups several standards for the achievement of which specific objectives (criteria) are set, which are achieved through concrete actions (requirements). Specifically, the criteria and requirements relating to hotel conditions are mentioned in Annex 1 of *the Standards for the Accreditation of Hospitals* of Order 446/2017.

Concerning the accommodation conditions for the *Strategic and Organizational Management* domain in standard 3, *human resources management ensures the staffing needs according to the mission assumed by the hospital. The third specific* objective is that *staffing needs are established according to the technical capacity, hotel, addressability, morbidity treated, staffing norms, and working time audit, as appropriate*. The first objective is that *the organization of the care environment* complies with the hospital's assumed capacity and competences and includes as a concrete action. *The hotel conditions respond to the particularities of each patient*. The second objective is that *the care environment is constantly assessed and adapted to the needs of the healthcare* and requires as a concrete action. *The institution constantly assesses and improves the hotel conditions*.

The way in which all these conditions must be met is described in the Accreditation Standards Manual for the Accreditation of Bed Health Facilities of the National Agency for Quality Management in Health Care (Cepoi & De Cecco, 2020). Thus, Standard 9 of the first domain analyzed at accreditation has a dual purpose. On the one hand, the health facility to set those conditions of the in-hospital care environment that can support the defense and regeneration of the patient's body and, at the same time, have an effect in terms of reducing as much as possible the harmful effects of stress generated by hospitalization, preoccupation with illness, disruption of links with work, home, family and friends, and psychological risks caused by the duration of hospitalization (anxiety, fear, loneliness, feelings of abandonment, alienation, helplessness), and, on the other hand, the hospital must provide the conditions that make up a safe care environment, with respect for cleanliness, disinfection, asepsis, and food safety, adapted to the patient and the comfort perceived by the patient or his or her carers (Cepoi & De Cecco, 2020).

To achieve these aims, Standard 9 of the Strategic and Organizational Management domain has three major objectives: (a) ensuring hotel and environmental conditions adapted to the services offered by the hospital and to the specific characteristics of the medical services offered to patients; (b) ensuring the cleanliness, disinfection, and antisepsis of all premises, equipment for a safe healthcare environment; and (c) providing food adapted to the patient in safe conditions. To accomplish these major objectives, hotel conditions will be supplemented with environmental conditions that provide a climate to enhance a patient's ability to recover: provision of green spaces with relaxation areas [intentionally or not this measure is in line with several studies who have suggested that manifestations of nature, such as a view of nature from a window, images of nature, and indoor plants, can positively affect patients' hospital experience through decreased pain, anxiety, and depression (Jamshidi et al., 2020)], providing an adapted noise threshold, lighting of individual rooms as well as common areas, use of appropriate colors, making adapted and marked trails for patients with physical particularities, providing an attendant during the movement of patients with special needs through the hospital, providing access to the hospital, wards, wards, wards, bathrooms, as well as furniture or bathrooms adapted to the particularities of patients or recreational facilities contribute to the formation of a climate conducive to the acquisition of health or balance, as well as to the comfort perceived by the patient or caregivers. Achieving the major objectives also requires ensuring cleanliness (disinfection and asepsis to avoid infections associated with the medical act and maintain patient safety) and a diet adapted to the patient (pathology of hospitalization, associated pathology, particular physiological states or preferences resulting as a consequence of religious beliefs, etc., safely provided as part of individualized care) (Cepoi & De Cecco, 2020).

The three major objectives, in turn, imply two specific objectives: (a) ensuring a care environment that meets the conditions of the hospital's assumed capacity and competencies and (b) organizing a care environment that is continuously assessed and adapted to the needs of healthcare.

The first specific objective of Standard 9 of the Strategic and Organizational Management domain is achieved according to the National Agency for Healthcare Quality Management Manual through the following concrete actions:

(a) In order to improve the quality of hotel services and their customization according to the patient, the health facility should provide: (1) hotel services to patients who do not require continuous hospitalization, at their request—conditions for hand washing and/or disinfection on entering and leaving each ward; (2) a functional and accessible alert system at the level of each sanitary group—individual and functional alert system for each bed; (3) beds-cots, without patient mobilization, for pathologies requiring this assessment; and (4) the implementation of these measures ensures a supportive climate with influence on a patient's ability to recover.
(b) Enforce rules that protect the dignity and privacy of the patient during the course of medical and surgical examinations, consultations, and treatment.
(c) The transportation of patients, in compliance with hospital rules, adapted to each service, on the one hand, by means of circuits and means of transport, with specific destinations, adequate in all respects and in sufficient numbers for all situations—the provision of stretchers and wheelchairs equipped with safety systems to ensure the patient's physical integrity during transportation, both in the patient reception area and in the wards/departments; and, on the other hand, by appointing staff to accompany the patient during movements within the hospital for the purposes of analysis and investigations.
(d) Carry out cleaning of the institution's premises and related equipment by implementing specific procedures and outlining responsibilities for ensuring cleanliness: compliance with the sanitization schedule for equipment and facilities; posting in a visible place the cleaning schedule at the level of the health unit premises; carrying out sanitization of each type of area/surface/operation with separately marked utensils, the hospital ensuring the necessary conditions for the performance of medical care; having at the level of each ward/compartment a functional ploscar for each care unit (25–30 beds); and different indication of elevators according to the specifics of transport.

(e) Definition of the circuits and assumption of the quality of sterilization sterilization at the hospital level by the application of the rules on how sterilization is carried out; implementation of appropriate measures to prevent contamination of personnel; sterilization service, that is, the use of protective equipment; and compilation of a nomenclature of medical instruments and devices, specifying the number of resterilizations allowed for each.
(f) Providing individualized medical care for each patient in safe conditions, by providing food adapted to the pathology for which the patient has been admitted, the associated pathology, particular physiological states or preferences resulting as a consequence of religious beliefs, and so on.
(g) The qualitative and quantitative provision of food for patients, relatives, and medical staff by the hospital, by: observance of the shelf life of products; proper storage of food and other materials; application of food quality control rules (organoleptic, quantitative, qualitative); use of protective equipment by kitchen staff; correlation of menus with ward requests (according to the pathology of hospitalized patients) by types of regimen; inclusion in the hospital's organizational chart of dietician posts by employment contract or civil agreement; possession of health cards by staff who prepare the food; epidemiological check of kitchen staff by an authorized person and recording it at the entrance of each shift; and maintenance of kitchen equipment.
(h) Respecting the hygiene circuits and rules at the food block level by ensuring for each ward a separate circuit for food handling; providing dedicated means of transport for transporting food and maintaining its temperature; designating staff to distribute food to patients; adopting specific rules governing the way of serving meals at the patient's bedside; and informing staff and patients about the menu and the food serving schedule.
(i) Enforcement of the rules on quality, hygiene, and linen circuit by: analyzing the quality of treatment of linen and effects; use of disposable linen in consultation rooms, treatment rooms and operating theaters, as the lack of linen and effects makes possible the transmission of diseases; establishment at the level of the health unit of indicators for monitoring the quality of the washing of linen and effects for patients, caregivers, and medical staff; operation at the level of the laundry service of the health unit of separate circuits for clean and soiled linen and effects; and equipping the hospital with pajamas, gowns, and slippers for patients.
(j) To prevent the occurrence of nosocomial infections by providing each ward with different colored containers for the selective collection of linen and hospital effects according to the probability of microbiological infestation; equipping with machines that allow automatic detection of microbial load; and using color coding to differentiate between clean and used linen. (Cepoi & De Cecco, 2020).

To achieve the second specific objective of Standard 9 of the Strategic and Organizational Management domains, the Manual of the National Agency for Quality Management in Health Care mentions the following concrete actions:

(a) The evaluation and continuous improvement of hotel conditions based on the analysis of data obtained from patient satisfaction questionnaires on accommodation and cleanliness—compliance with cleanliness, disinfection, and asepsis protects the patient from healthcare-associated infections and contributes to maintaining patient safety.
(b) The evaluation and improvement of food supply and storage services, as well as food preparation services, taking into account the analysis of data collected from: patient satisfaction questionnaires and patient and/or staff complaints/complaints about the food received—food adapted to the pathology for which the patient has been admitted, the associated pathology, particular physiological states or preferences resulting from religious beliefs, and so on, provided in safe conditions, is part of the individualized care provided to the patient.
(c) The constant analysis and improvement of the laundry service following the processing of information obtained either from patient satisfaction questionnaires or from patient and/or staff complaints about the quality of linen and linen effects—regular replacement of linen and linen effects and the use of disposable linen is necessary to maintain proper hygiene and, implicitly, to prevent the risk of healthcare-associated infections.
(d) The provision of a suitable indoor and outdoor environment that includes: appropriate room temperatures, monitored and recorded according to the specifics of the service; furniture in waiting areas (chairs, armchairs, sofas, or benches) and furniture or bathrooms adapted to the specifics of the patients); spaces for recreation of patients and their relatives; and the provision of access to the hospital, wards, wards, and bathrooms adapted to the specifics of the patients (Cepoi & De Cecco, 2020).

Concrete evaluation of hotel service delivery, and especially the impact on the patient, is very difficult to realize. In a not-so-recent study, not long after the adoption of Law No. 95/2006, focused on the situation in psychiatric hospitals, it was concluded that: "Housing conditions are also relatively low for current standards, with little concern for the privacy and environment of the patient" (Junjan et al., 2009). But this is a regional problem because patient experience is low in Hungary as well (Zsarnoczky-Dulhazi et al., 2023).

Now, the Order 1501/2016 *on the approval of the implementation of the patient feedback mechanism in public hospitals* includes in Annex 1 a form in which the Minister of Health invites discharged patients to answer anonymously and free of charge 8 questions in order to improve health services, but none of these questions refer to the accommodation conditions. However, the questionnaires in Annex 65 and 66 of the *Order of the President of the National Agency for Quality Management in Health Care No. 8/18.01.2018* also include queries on some hotel issues. Thus, in Annex 65 entitled *Patient Interview*, question 4 refers to whether the patient was provided with a locker or wardrobe where to store the clothes and personal belongings with which he/she came, and the next question asks whether the patient was provided with a set of pajamas, robes, slippers, matching, and bed linen from the hospital. Also, question 11 refers to the existence of a restroom in the ward, and the next two questions concern the information about the meal program and the menu of the day. In Annex 66 entitled *Patient Questionnaire*, question 6 refers to cleanliness in the hospital lounge.

In addition to the standard obligatory hotel conditions to be provided to all patients, hospitals may additionally offer those who request two other categories of services: (a) superior hotel services and (b) high comfort hotel services, respectively. However, since these conditions exceed the generally accepted needs, the law only stipulates that their cost will not be covered by the public health insurance system, but in the case of superior hotel services, by the private insurer according to Article 349 para. (3) of Law No. 95/2006 based on the health insurance contract, respectively, by the patient himself for the high comfort hotel services according to Article 248 para. (1) letter d of Law No. 95/2006.

Hospitals may offer hotel services not only to patients but also to other persons. This possibility is expressly provided for by Article 193 para. (7) of Law No 95/2006: "Public hospitals may realize additional income from: ... g) medical, hotel or other services provided at the request of third parties." However, the presence of these persons staying in the hospital should represent a benefit for the patient. The idea is supported by the provisions of Article 233 of Law No. 95/2006: "The costs of hotel services for the person accompanying a hospitalized child up to the age of 3 and for the companion of a severely disabled person hospitalized shall be borne by the health insurance funds, if the doctor deems their presence necessary for a specific period. The cost of hotel services for a person accompanying a child over 3 years of age admitted to a public hospital, borne by the accompanying person, may not exceed twice the amount of the food allowance specific to the ward in which the child is admitted. In other words, the beneficiaries of the hotel services provided by the hospital must contribute to the well-being of the patients hospitalized in that hospital, and, if their presence is considered necessary by the doctor for the provision of medical care, the cost of the hotel service is covered for a determined period by the public health insurance system."

First, in the case of minor patients, there is an objective need for their parents or guardians to be accommodated in the hospital with them. Starting even with newborns, according to Annex 1 *Hospital Accreditation Standards* of Order 446/2017, domain 2 *Clinical Management*, the fourth standard is "*The hospital promotes the concept of"child-friendly""* for which the specific objective *the hospital has adopted a policy to promote breastfeeding in neonatology wards is realized by the hospital through a concrete action whereby hospitalized mothers are informed about the benefits of breastfeeding*. Consequently, mothers of newborn patients are accommodated in the hospital in order to be able to breastfeed their newborns. Also, for other minor patients, the presence of parents may be useful for the psychological comfort of the patient or to mediate communication with the medical staff, but it is certainly legally necessary to obtain the consent of the legal representative to the provision of the medical service. Thus, pursuant to Article 661 of Law No. 95/2006, this means that all patients have the right to medical consent, but minor patients do not have the legal capacity to exercise this right, and, consequently, it is expressed only in written form by the legal representative (parents or guardian)—hence, the need for the presence of the legal representative not only for the limited period of the patients' visiting hours in the hospital but also for a longer period of time, which can be facilitated by the accommodation of the parent or guardian in the hospital.

Second, in the case of severely disabled patients, there is an objective need to be assisted by their attendant. Thus, according to Article 2 para. (1) of Law No. 448/2006 *on the protection and promotion of the rights of persons with disabilities*: "Persons with disabilities are those persons whose social environment, not adapted to their physical, sensory, mental, psychological, mental and/or associated impairments, totally prevents or limits their access with equal opportunities to the life of society, requiring protective measures to support their social integration and inclusion." To this end, persons with disabilities may have an attendant, and persons with severe disabilities a personal assistant defined by Article 5, point 6 of Law No. 448/2006 as the person "who supervises, assists and provides care to the child or adult with severe disability, based on the habilitation-rehabilitation plan for the child with disability, respectively the individual service plan of the adult with disability". Precisely because the individual plan must also be followed during hospitalization for other conditions, one of the rights of the disabled patient provided by Article 11 paragraph (1) letter b of Law No. 448/2006 is: free accommodation and meals and for the accompanying person of the severely or severely disabled child or of the severely or severely disabled adult in health facilities with beds, sanatoriums, and spas, on the recommendation of the family doctor or specialist, provided from the National Single Fund for Social Health Insurance, according to the Framework Contract on the conditions of healthcare in the social health insurance system.

For these hotel services, the law does not provide for special liability. Therefore, a question to be analyzed is raised by the provisions of Article 42 para. (3) of the Annex to the Order of the Minister of National Defense No. M 110/2009 for the approval of the Instructions on medical and pharmaceutical assistance in the Ministry of National Defense in peacetime, According to it, in military hospitals: "The hospital commander, the head of the ward and the attending physician are responsible, under the conditions of the law, for the quality of the medical act, the compliance with the conditions of accommodation, hygiene, nutrition and prevention of nosocomial infections, as well as for the coverage of the damages caused to the patients." Thus, we are clearly in front of a legal innovation of medical malpractice for accommodation and food. Medical civil liability in general, and the civil liability of the doctor in particular, is not limited to civil liability for medical malpractice situations because there are situations in which the doctor is civilly liable in his capacity as a doctor, but not for the manner in which he rendered the medical service (e.g., in the hypothesis that a donation or a legacy is annulled for violation of Article 990 Civil Cod). But the problem is that Article 42 para. (3) of the Annex to Order No. M 110/2009 establishes as situations of malpractice (not explicitly, but by inserting in the list of situations of malpractice that begins with the quality of the medical act and ends with nosocomial infections) the damage caused by the conditions of accommodation and food (Vida-Simiti, 2018).

First, the provisions of Articles 653–654 of Law No. 95/2006 are explicit and limitative by listing the situations in which the doctor is obliged to repair the patient's damage caused by the medical service rendered defectively due to negligence or imprudence, lack of informed consent, lack of medical knowledge or exceeding professional competence, and the extension of medical malpractice situations to the compensation of the damage suffered by the patient as a result of the accommodation and food conditions exceeds these legal limits.

Second, Romanian law does not make an exact distinction between the healthcare contract and the hotel contract, but this does not mean that, in the content of the medical contract, the obligation to provide the medical service is confused with the obligation to provide hotel services. On the contrary, the distinction is made for the public health system by Article 193 para. (7) of Law No. 95/2006: "Public hospitals may earn additional income from: ... g) medical, hotel or other services provided at the request of third parties"; and for the private health system by Article 349 para. (3) of Law No. 95/2006: "Voluntary supplementary health insurance covers all or part of the payment for any type of services not included in the basic medical services package, the option for a specific medical staff, the request for a second medical opinion, superior hotel conditions, other medical services specified in the insurance policy." Moreover, it follows from the economy of Article 248 of Law No. 95/2006 that hotel services are not always an accessory of the medical service, but, on the contrary, distinct contractual obligations whose existence and extent do not depend on the medical services: "The services that are not paid from the fund, their value being borne by the insured, by the requesting units, from the state budget or from other sources, as appropriate, are: ... d) hotel services with a high degree of comfort ... p) hotel services requested by patients whose illnesses are treated in day hospitalization." Precisely in this sense, according to Article 233 of Law No. 95/2006, the actual beneficiary of the hotel service may also be a person distinct from the beneficiary of the medical service: "The hotel service charges for the person accompanying the hospitalized child under 3 years of age, as well as for the accompanying person of the hospitalized severely disabled person, shall be borne by the insurance companies, if the doctor deems their presence necessary for a specific period."

In principle, hotel obligations are obligations to give, that is, to cede the use (in whole or in part) of immovable property (a room or part thereof) and/or movable accessory goods (bed, wardrobe, bedside table, television, towels, etc.—but not appliances, equipment, medical devices, which are ancillary to the medical service) in exchange for a price (the creditor of the obligation to pay the price is the hospital, under Article 193 para. (3), and the debtor is the patient who pays the price directly, pursuant to Article 248, or indirectly through the insurer, pursuant to Article 349 para. (3) of Law 95/2006. For this reason, the clauses in the medical contract (to which only the hospital can be a contracting party since, pursuant to Article 163 of Law 95/2006, it is the sole provider of medical services with beds) relating to these obligations are similar to a lease (Articles 1777–1823 Civil Code). The main obligations of the hospital-landlord under Article 1786 are: (1) to hand over the main and ancillary goods which are the subject of the hotel clause in the medical contract; (2) to maintain the goods in a proper state of use (which implies, due to the nature of the contract in which the hotel clause is included, in addition to repairs—an inherent obligation of the landlord under Article 1788 Civil Code—the provision of hygiene and the facilities required by the patient's various permanent or temporary disabilities); and (3) to ensure that the tenant-patient is assured of the quiet and useful use of the property. The hospital-landlord is civilly liable for the non-fulfillment lato sensu of these obligations, including for defects of the good and lack of quality.

Although not as detailed as in common law, the hospital's obligation regarding hotel services is also laid down by special law in Article 166 para. (1) of Law No. 95/2006: "The hospital shall ensure conditions for medical investigations, treatment, accommodation, hygiene, food and the prevention of nosocomial infections, in accordance with the rules approved by order of the Minister of Health," which is not a mere project of intention, but is the basis of a precise liability provided for by Article 166 paragraph (2) of Law No. 95/2006: "The hospital is responsible, under the conditions of the law, for the quality of the medical act, for the respect of the conditions of accommodation, hygiene, food and prevention of nosocomial infections, as well as for the coverage of the damages caused to the patients." Moreover, Article 42 para. (2) of the Annex to Order No. 110/2009 ("The hospital shall ensure conditions for clinical, paraclinical, and laboratory examinations, medical and surgical treatment, recovery, care, medical evacuation, medicines and sanitary materials, accommodation, hygiene, food and prevention of nosocomial infections") reproduces ad litteram the legal provisions of Article 166 para. (1) of Law No. 95/2006; however, in the following paragraph, it exempts the debtor-hospital from liability by holding the commander, the head of the department, and the attending physician responsible for the failure to perform the obligation in the broad sense of the word.

12.5 Conclusions

A first conclusion is that, from the legal point of view merger of the two, the healthcare industry and the hospitality industry, has already happened because, based on Article 166 para. (1) of Law No. 95/2006, a hospital "is a health facility with beds," which means that hotel services are part of the essence of the hospital as a health unit. That is why the criteria and requirements relating to hotel conditions for the assessment of hospitals are mentioned in Annex 1 of *the Standards for the Accreditation of Hospitals* of Order 446/2017 *on the approval of the Standards, Procedure, and Methodology for the Assessment and Accreditation of Hospitals* and the way in which all these conditions must be met is described in the Accreditation Standards Manual for the Accreditation of Bed Health Facilities of the National Agency for Quality Management in Health Care. So, hospitals are required by law to improve patient healthcare experience, but the problem is the evaluation of hotel service delivery and especially the impact on the patient is very difficult to realize because, on the one hand, the Order 1501/2016 *on the approval of the implementation of the patient feedback mechanism in public hospitals* includes in Annex 1 a form in which discharged patients are invited to answer questions in order to improve health services, none of which refers to the accommodation conditions, and, on the other hand, the questionnaires in Annex 65 and 66 of the *Order of the President of the National Agency for Quality Management in Health Care No. 8/18.01.2018* include only very general queries on some hotel issues such as the existence of a locker or wardrobe where to store the clothes and personal belongings with which he/she came, or the provision of a set of pajamas, robes, slippers, matching, and bed linen from the hospital.

However, another conclusion is that the legal obligation of hospitals (public and private) to improve the customer experience can also be a business opportunity because the Romanian hospitals can provide three kinds of hotel services: (a) basic hotel services (for which the cost will be covered by the public health insurance system), (b) superior hotel services (with the cost covered by a private insurer according to Article 349 para. 3 of the Law No. 95/2006 on the basis of the health insurance contract), and (c) high comfort hotel services (paid by the patient himself according to Article 248 para.1 letter d of the Law No. 95/2006). More than that, these services may be provided to both patients and third parties. However, the beneficiaries of the hotel services provided by the hospital must contribute to the welfare of the patients hospitalized in that hospital, which is why Article 233 of Law No. 95/2006 expressly refers to the mothers of underage patients and, respectively, the accompanying persons of severely disabled persons.

The third conclusion regards the legal liability. Article 166 para. (2) of Law No. 95/2006 establishes the hospital's responsibility for these hotel services, but without mentioning a special legal liability. That is why, in the case of military hospitals, the Ministry of Defense has tried to transfer through Article 42 para. (3) of the Annex to Order No. M 110/2009, the responsibility for the accommodation and food conditions from the hospital to the medical staff as a new form a malpractice liability. But this is a breach of the law. First, the provisions of Articles 653 and 654 of Law No. 95/2006 are explicit and limitative by listing the situations in which the doctor is obliged to repair the patient's damage caused by the medical service rendered defectively due to negligence or imprudence, lack of informed consent, lack of medical knowledge or exceeding professional competence, and the extension of medical malpractice situations to the compensation of the damage suffered by the patient as a result of the accommodation and food conditions exceeds these legal limits. Second, because the hotel services are not always an accessory of the medical service, but, on the contrary, distinct contractual obligations to give, that is, to cede the use (in whole or in part) of immovable property (a room or part thereof) and/or movable accessory goods (bed, wardrobe, bedside table, television, towels, etc.—but not appliances, equipment, and medical devices, which are ancillary to the medical service) in exchange for a price (the creditor of the obligation to pay the price is the hospital). For this reason, the clauses in the medical contract relating to these obligations are similar to a lease (Articles 1777–1823 Civil Code). The hospital-landlord is civilly liable for the non-fulfillment lato sensu of these obligations, including for defects of the good and lack of quality.

References

Cepoi V., & De Cecco N. (2020). *Accreditation standards manual for the accreditation of bed health facilities*. Accreditation Cluster II, Bucharest, National Agency for Quality Management in Health Care. https://anmcs.gov.ro/web/wp-content/uploads/2021/02/Manualul-standardelor-de-acreditare-2020.pdf. Accessed 12 Apr 2025.

Commission Regulation (EU) 2021/1901 *of 29 October 2021 implementing Regulation (EC) No 1338/2008 of the European Parliament and of the Council as regards statistics on health care expenditure and financing* published in the Official Journal of the European Union L 387 of 3.11.2021.

Dorsner-Dolivet, A. (2006). *La Responsabilité du Médicine*. Economica.

Jamshidi, S., Parker, J. S., & Hashemi, S. (2020). The effects of environmental factors on the patient outcomes in hospital environments: A review of literature. *Frontiers of Architectural Research, 9*(2020), 249–263.

Junjan, V., Miculiţa, I., Popescu, C., Ciumăgeanu, M., Sfetcu, R., & nGhenea, D. (2009). The provisions of psihychiatric care in Romania – Need for change or change of need? *Transylvanian Review of Administrative Sciences, 26E/2009*, 81–93.

Law no. 95/2006 *on Health Care Reform,* published in the Official Journal of Romania no.652 of 28.08.2015.

Law no. 448/2006 *on the protection and promotion of the rights of persons with disabilities* published in the Official Journal of Romania no. 1 of. 3.01.2008.

Majeed, S., & Kim, W. G. (2023). Toward understanding healthcare hospitality and the antecedents and outcomes of patient-guest hospital-hotel choice decisions: A scoping review. *International Journal of Hospitality Management, 112*, 103383. https://doi.org/10.1016/j.ijhm.2022.103383

Order of the Minister of National Defense No. M 110/2009 *for the approval of the Instructions on medical and pharmaceutical assistance in the Ministry of National Defense in peacetime* published in the Official Journal of Romania no.735 of. 29.20.2009.

Order of Ministry of Health no 1501/2016 *on the approval of the implementation of the patient feedback mechanism in public hospitals* published in the Official Journal of Romania no. 18 of. 9.01.2017.

Order of Ministry of Health no. 446/2017 *on the approval of the Standards, Procedure and Methodology for the Assessment and Accreditation of Hospitals* published in the Official Journal of Romania no. 300 of 27.04.2017.

Regulation (EC) No 1338/2008 *of The European Parliament and of the Council of 16 December 2008 on Community statistics on public health and health and safety at work* published in the Official Journal of the European Union L 354 of 31.12.2008.

Vida-Simiti, I. (2013). *Civil liability of the doctor*. Hamangiu.

Vida-Simiti, I. (2018). Military doctor's liability for medical malpractice aggravated beyond the limits of the law. *Dreptul, 1*, 65–73.

Zsarnoczky-Dulhazi, F., Zsarnoczky, M., Kopeer, B., Karpati, J., Molnar, C., Collins Adol, G. F., & Denes David, L. (2023). Promising European research results to improve hospitality in healthcare by eHealth. *International Journal of Hospitality Management, 112*(2023), 103411. https://doi.org/10.1016/j.ijhm.2022.103411

Chapter 13
Critical Review of Customer Experience Research in Sports Events

Oana Ruxandra Bode and Adina Letiţia Negruşa

Abstract One of the physical activities in our daily life takes the form of sports. Frequently characterized by competition and structure, sports engage, sustain, or enhance physical capabilities and talents. Participants also find them enjoyable, and in certain instances, they offer entertainment for spectators. Sports events are defined as those that are organized according to a specific predetermined program, have a significant social and economic impact on the location or region where they are held, and are distinguished by creative and complex content of sports-like, recreational activities of an entertaining character. This study is mainly focused on investigating emerging themes and identifying new trends in the study of Customer Experience and Tourist Behaviour (CETB) in relation to sports events. It encompasses science mapping and an in-depth examination of thematic evolution in subtopics, citation trends, co-occurrence networks, and development paths within CETB research related to sports events. The Scopus database was chosen as the main source of bibliometric analysis for this research, since it ensures an accurate and detailed exploration of the research work, facilitating the discovery of critical insights into the thematic and structural development of academic inquiry. The software VOS viewer is used to create and display bibliometric networks, and 183 articles distributed across 81 distinct journals were considered useful for the study. One more objective of the analysis is to generate a VOS viewer diagram, which contributes to a clearer understanding of the intellectual structure of the field, guiding future research directions and practical applications. By reviewing sports event studies in the international literature, this research seeks to close a gap in the area and give academics a comprehensive viewpoint. It helps to clarify the intellectual

O. R. Bode (✉)
Faculty of Business, Department of Hospitality Services, Babeş-Bolyai University, Cluj-Napoca, Romania
e-mail: oana.bode@ubbcluj.ro

A. L. Negruşa
Faculty of Business, Department of Hospitality Services, Babeş-Bolyai University, Cluj-Napoca, Romania

A. L. Negrusa, M. M. Coroş (eds.), *Innovation, Sustainability, and Growth in a VUCA Environment*, Springer Proceedings in Business and Economics,
https://doi.org/10.1007/978-3-032-11639-0_13

framework of the field, and, using the idea of co-occurrence, pinpoint academic hotspots and major research topics.

Keywords Sports events · Tourism · Customer experience · VOS viewer diagram

JEL Classification L83 · L80 · Z29 · Z32

13.1 Introduction

Sports engage, maintain, or improve physical abilities and talents and are often defined by structure and competition. It offers both participants and spectators an exciting and energizing experience. Engaging in sports fosters a spirit of sportsmanship that extends beyond the game itself and influences our approach to life. Also, it provides a thrilling and energetic experience for both those involved and those watching. Investments in the sports industry during the 1970s contributed to the rise in sports awareness and habits around the world, respectively, to the recognition of sports as an essential aspect of life, and to people's desire to continue their regular sports routines while traveling.

Over the years, it was revealed that two key factors in the promotion and long-term economic growth of tourist destinations are sport and tourism. In this way, sports and tourism became complementary. Therefore, sport tourism has become a significant component of the global tourism industry. Traveling to take part in an active (as participant) or passive (as spectator) sport-oriented holiday is known as sport tourism. United Nations Tourism defines sports tourism as "a type of tourism activity which refers to the travel experience of the tourist who either observes as a spectator or actively participates in a sporting event generally involving commercial and non-commercial activities of a competitive nature."

The main drivers of the sport tourism industry's expansion over the past years have been people's increased free time, disposable income, affordable travel options, and, most importantly, the increased interest in sports events around the world. With approximately 10% of global tourism spending attributed to sports tourism, it has emerged as a major economic driver, moving masses across continents and demonstrating an anticipated growth rate of 17.5% for the period 2023–2030 (United Nations World Tourism Organization). Sports tourism has the potential to further development, foster social, economic, and environmental action, and leave a lasting beneficial impact. Thus, the 2010 Sport Tourism Expo brought to light the interconnectedness of sport, sporting events, travel, and tourism for the first time. Sports events are therefore at the intersection of numerous academic fields, requiring an interdisciplinary approach.

As a result, it was found that the tourist industry segment with the fastest rate of growth is travel associated with sports or physical activities, and sports event tourism has drawn considerable academic attention, becoming a prominent and thoroughly studied subsectors of the tourism area (Wäsche et al., 2013). Primarily, the studies related to sports events concentrate on sustainability, economic impact, and destination image. Numerous studies have examined the relationship between sports events and tourism, on one hand, as well as their socioeconomic implications, on the other hand. The potential of hosting sporting events to support the socio-economic development of towns and regions has been the subject of numerous studies (Chalip & Leyns, 2002; Vogt & Andereck, 2003; Fourie & Santana-Gallego, 2011; Liu, 2015). Additionally, in the increasingly competitive globalized world, sports event tourism can help towns and regions stand out from the competition (Kolb, 2006; Pike, 2013). Furthermore, researchers emphasized the reasons for attending to sporting events and connected these to individual demographic characteristics as gender, age, place of residence, socioeconomic status, family and group affiliation, self-esteem, knowledge, and social interaction. This was essential since these reasons for traveling enhance one's awareness of the host nation's culture, tourism attractions, and general destination offerings. While these studies provide valuable insights, one crucial aspect emerges for further exploration: customer experience (CX) in sports events.

To ensure the consistent interpretation of the research terminology, first, standardized definitions of core terms are provided, thereby offering essential context for understanding the study's scope and approach. Initially, CX represented a straightforward concept without the multidimensional complexity that characterizes people's engagement with specific activities. According to the literature reviewed by Jain et al. (2017), CX is defined as: "the aggregate of feelings, perceptions and attitudes formed during the entire process of decision making and consumption chain involving an integrated series of interaction with people, objects, processes and environment, leading to cognitive, emotional, sensorial and behavioural responses." A CX framework integrating digital, social, and physical realities was proposed by Bolton et al. (2018), who defined the concept as: "encompassing customers' cognitive, emotional, social, sensory and value responses to the organization's offerings over time, including pre- and post-consumption." Therefore, the research revealed that customer interactions should occur through touchpoints—critical moments when customers exhibit cognitive, behavioural, emotional, social, and physical responses. Limited attempts have been made to define CX within the sports context as well. Theodorakis (2014) examined CX in sports events as a multidimensional concept, resulting in the development of a conceptual model: "holistic in nature and involves the spectator's cognitive, affective, emotional, social, and physical responses to the sports event and its organizer (i.e., the sporting team)." The definition given by Shtengelov (2020), "Emotional, cognitive, behavioral, sensorial and social responses given by the spectators on interactions between the spectators and sport organizations at the venue or via online platforms before, during and after a sport event," acknowledges the dynamic nature of lived experiences, as emotional, cognitive, behavioural, sensorial, and social responses vary according to multiple factors including team performance, facility conditions, service quality, interactions with other customers, and organizational elements.

CX is a multidimensional concept encompassing cognitive, emotional, physical, sensorial, spiritual, and social elements that shape a customer's direct or indirect interactions with market actors (De Keyser et al., 2015). At its core, it involves all individual encounters between an organization and its customers across various stages, known as touchpoints (Homburg et al., 2017). Both scholars and practitioners emphasize the importance of identifying, visualizing, analysing, and improving these touchpoints from the customer's perspective to enhance the overall experience (Blomkvist & Segelström, 2014). Lemon and Verhoef (2016) conceptualize CX as a dynamic journey, unfolding across multiple touchpoints throughout the entire purchase cycle. This journey-oriented view offers a deeper understanding of how customers navigate their options and make choices at each stage, from initial consideration to post-purchase engagement.

In the hospitality and tourism sectors, the concept of CX has undergone significant evolution, becoming a critical determinant of both consumer behaviour and organizational success. It offers a comprehensive viewpoint on customer service interactions, highlighting the varying subjective emotions and emotional reactions that clients go through (Chepngetich et al., 2019). This shift highlights a move beyond traditional service quality metrics towards a deeper, more nuanced understanding of customer satisfaction. Consequently, adopting a customer-oriented approach has become essential, especially in increasingly competitive and experience-driven markets where consumer preferences are continuously evolving. Beyond their broader impact on tourism, sports events are also deeply rooted in individual attendee experiences. The way the audience interact with an event, from their emotional connection to a team to the social atmosphere within the venue, plays a fundamental role in shaping perceptions of sports tourism and influencing future participation. Therefore, CX in sports events extends beyond mere attendance, encompassing emotional, social, and sensory dimensions that contribute to an engaging and memorable interaction. Scholars highlight those key environmental factors, such as facility quality, stadium ambiance, and accessibility, that serve as fundamental drivers of attendees' satisfaction (Wakefield & Blodgett, 1996). The emotional aspect of CX is equally important, as fans connect deeply with teams and athletes, forming lasting memories and reinforcing brand loyalty (Theodorakis et al., 2016). Social interaction plays a critical role as well, with shared event experiences fostering a strong sense of belonging among attendees (Funk & James, 2001).

Nevertheless, challenges still exist; thus, this study examines the current state of research on CX at sports events, identifying significant themes, methodologies, and areas in need of further investigation. By synthesizing previous findings, it aims to highlight consumer motivations, satisfaction drivers, and behavioural trends in sports event tourism. The insights gained will inform event organizers, tourism stakeholders, and policymakers, contributing to strategies that enhance the overall sports event experience while increasing destination appeal.

13.2 Context of the Study and Research Methodology

Events are meticulously designed experiences that shape the interactions of both attendees and participants, serving diverse objectives across social, artistic, entertainment, educational, and cultural domains. Within this broad spectrum, sports events occupy a unique space, characterized by their structured format, wide-ranging appeal, a significant social and economic impact on the location or region where they are held, and a distinguished content of sports-like, recreational activities of an entertaining character.

Sports events have evolved significantly throughout history. Initially emerging as ceremonial public gatherings linked to religious traditions, their existence as organized spectacles can be traced back to at least the first Olympic Games in 776 BC (Matheson, 2006). Today, sports events range from local competitions to globally recognized tournaments, offering destinations an opportunity to boost their tourism appeal. Research suggests that sports events are among the most influential tourist attractions, driving visitor numbers and enhancing local economies (Kaplanidou & Vogt, 2007). As a result, sports events are regarded as a central pillar of sports tourism, occurring in various formats and scales (Kaplanidou & Vogt, 2010).

Beyond their role in tourism promotion, sports events are also deeply intertwined with CX. Attendees do not merely observe an event; they actively shape their own engagement through emotional connections, social interactions, and sensory impressions. This dynamic interaction reinforces the importance of analysing CX in sports events, as it plays a crucial role in event success, visitor retention, and destination appeal.

To address the main objective of this study, analysing existing literature on CX in sports events, two methodological approaches were employed:

A systematic bibliometric review of journal publications spanning from 2000 to 2025 was conducted. This approach enabled the identification of key research trends, influential studies, and evolving themes in sports event tourism. By mapping scholarly output and citation patterns, the bibliometric analysis provided insight into the progression of CX research within sports events, offering a data-driven foundation for further investigation.

A structured review of the most relevant studies was conducted to examine existing theoretical frameworks, research methodologies, and empirical findings. This analysis explored spectator satisfaction, emotional engagement, social dynamics, and technological innovations influencing CX. Furthermore, the review identified gaps in the current research landscape, such as the impact of cultural differences on event perception, strategies for sustaining post-event engagement, and the integration of sustainable practices in enhancing spectator experience.

Together, these methodological approaches provide a comprehensive framework for evaluating CX in sports events. By synthesizing insights from previous studies, this research aims to contribute to the ongoing discourse in sports event tourism, offering practical recommendations for event organizers, tourism stakeholders, and policymakers seeking to enhance the overall spectator experience and maximize event impact.

13.3 Results and Discussions

The Bibliometric Analysis

Bibliometric analysis serves as a powerful tool for examining the body of literature in a specific subject area by quantifying and comparing research to reveal its structural knowledge and thematic patterns (Okubo, 1997; Zhang et al., 2020). This study encompasses science mapping and an in-depth examination of thematic evolution in subtopics, citation trends, co-occurrence networks, and development paths within Customer Experience and Tourist Behaviour (CETB) research related to sports events.

Bibliometric tools have proven to be highly valuable for analysing and comprehending the scope and depth of academic research within various disciplines. These tools enable a systematic assessment of research publications compiled in a selected database over a defined period, evaluating their impact on the existing body of knowledge. Such an analysis offers insights into the scope of the subject, the progression of research in both its central and peripheral areas, as well as underlying patterns and emerging trends (Rai & Singh, 2020).

Using bibliometric methodologies, researchers can demarcate the body of scholarly work, quantitatively measure publication units, citation patterns, and overall research output (Broadus, 1987). These techniques have become instrumental across a wide range of academic fields, providing a quantitative framework for analysing, describing, and visualizing the literature landscape (Zhang et al., 2020). By systematically identifying impactful studies, high-performing authors, and influential journals, bibliometrics enhances the effectiveness of literature reviews. For the current study, the Scopus database (accessible at https://www.scopus.com/search) was selected as the primary source of bibliometric analysis. This choice was motivated by Scopus's standing as a globally recognized citation database, offering access to an extensive repository of scholarly work that is highly regarded for its reliability, comprehensive scope, and quality of publications. As a foundation of the systematic review process, Scopus follows rigorous criteria that align with the needs of this study. Moreover, its powerful indexing tool and advanced search features ensure an accurate and detailed exploration of the research work, facilitating the discovery of critical insights into the thematic and structural development of academic inquiry.

Review Approach

The authors conducted an advanced search in the Scopus database using the keyword "SE = (Sports, Sport, Event, Events)" to identify relevant studies in this research area. This initial search generated a total of 693 records on April 20, 2025. Subsequently, the search was further refined by incorporating the keywords "Customer Experience and Tourist Behaviour," which narrowed the results to 365. Furthermore, filtering was applied by narrowing the focus to publications categorized under Sports Events from 2000 to 2025, with conference abstracts and book reviews excluded from consideration. This refinement resulted in 228 peer-reviewed articles suitable for in-depth scholarly analysis in the chosen area. In the fourth phase, a language filter was applied, limiting the selection to articles published in English, which finalized the dataset to 225 records considered useful for the study.

Data Sources

In the final phase, the titles and abstracts of each article were rigorously reviewed to ensure both accuracy and relevance (X. W. Su et al., 2019). This process resulted in the selection of 183 records after excluding conference proceedings and articles that were not aligned with the research focus. This dataset was subsequently imported into the VOS viewer software, enabling systematic data extraction and detailed analysis.

The analysis provided a detailed exploration of the publications, focusing on critical elements such as sources, authors, documents, conceptual frameworks, and social structures within the study's domain. The findings are presented in a structured manner, combining performance analysis and science mapping to address key dimensions, including research publication growth patterns, leading publication sources, highly influential authors, cited reference trends, keyword dynamics, thematic evolution, and co-occurrence networks. These indices are widely recognized as robust tools for assessing academic research output and revealing the thematic dimensions of scholarly work. Bibliometric methodologies formed the backbone of this analysis. Performance analysis highlighted the contributions of authors, publications, and institutions (Zupic & Čater, 2015), while science mapping provided visual representations of intellectual connections within the field. These mappings captured a static snapshot of the field's structural framework at a given point in time (Small, 1973; Zupic & Čater, 2015), offering valuable insights into its development and interrelationships.

Data Analysis

The initial part of the analysis centred on tracing the evolution of publications in the field of sports events and tourist/CX from 2000 to 2025, alongside the number of citations these works garnered. As shown in Fig. 13.1, interest in these topics started

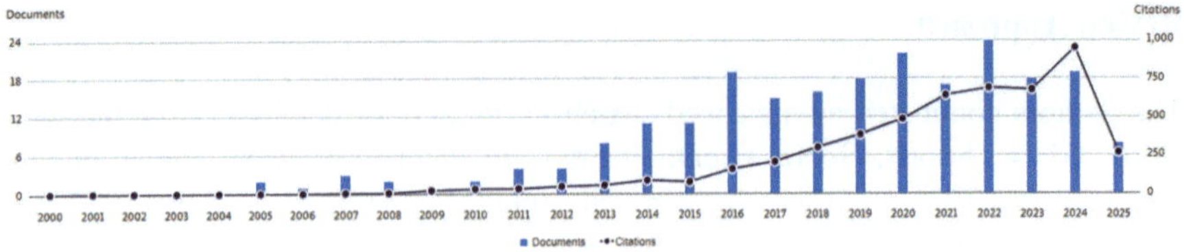

Fig. 13.1 Evolution of publications and citations metrics during the period 2000–2025. (Source: Authors' own processing)

Fig. 13.2 Top of journals according to the number of publications. (Source: Authors' own processing)

to accelerate prominently in 2014. Between 2019 and 2024, academic interest in this area experienced a significant growth, with an annual average of 19 publications, setting its importance as a key focus of academic research. In contrast, prior to 2014, the annual average of publications remained below five. This notable growth reflects the rising societal appeal of sports events, which has intensified the scientific inquiry into their promotion, organizational dynamics, and wider socio-economic impacts.

The analysis of publications and scientific journals during the reviewed period highlights a notable dispersion of research across multiple outlets, rather than a strong concentration within a select group of journals. As shown in Fig. 13.2, the top seven journals, which published the highest number of articles in this area, collectively account for only 38% of the total publications published between 2000 and 2025. This indicates that no single journal or small group of journals dominates the academic discourse on sports events.

Furthermore, it is worth emphasizing that the 183 analysed articles are distributed across 81 distinct journals. This wide-ranging dissemination reflects the multidisciplinary nature of research on sports events, encompassing diverse perspectives. The breadth of publication venues suggests that scholars approach this topic from a variety of angles, contributing to its dynamic and evolving landscape (Table 13.1).

Table 13.1 Top of journals according to the number of publications

Journals	Total no of publications for 2000–2025
Sustainability Switzerland	21
Tourism Management	12
Journal of Sport and Tourism	9
Event Management	8
Sport in Society	8
Leisure Studies	6
International Journal of Event and Festival Management	6
Journal of Sports Management	2

Source: Authors' own processing

The diversity observed in publication outlets highlights the growing importance of sports events as a research area. Such a distributed publishing pattern is essential for fostering innovation and ensuring that emerging insights reach a wider audience, stimulating further exploration of this area of study.

Trends and Resulting Clusters from the Overlay Visualization

The second part of the analysis aimed to identify predominant research areas and academic hotspots by utilizing the concept of co-occurrence (Morris & Van der Veer Martens, 2008). This method is rooted in the bibliometric practice and focuses on uncovering the relationships between keywords frequently used in scientific publications. The study relied on author keywords attached to main papers, together with the Keywords Index extracted using a computer algorithm designed to capture the essence of its abstract and the article's content. These keywords formed the pillar for generating the VOS viewer diagram illustrated in Fig. 13.3.

The resulting visualization map comprises 46 unique keywords, each appearing at least ten times in the dataset, distributed across four distinct clusters. Additionally, the map is characterized by 443 links and a total link strength of 1038, indicating the interconnectedness and relevance of the identified terms within the research domain (Fig. 13.4).

The clusters were systematically named by the authors based on the dominant keywords appearing within each group, as outlined below:

Cluster 1. (Magenta) *Customer Experience and Satisfaction in Sports Events*
Cluster 2. (Green) *Mega-Events and Sustainability*
Cluster 3. (Yellow) *Destination Image and Marketing in Sports Tourism and Events*
Cluster 4. (Blue) *Demographic and Behavioural Research in Sports Events*

The visualization map provides a dynamic representation of the interconnections among these clusters, offering valuable insights into how these themes interact and

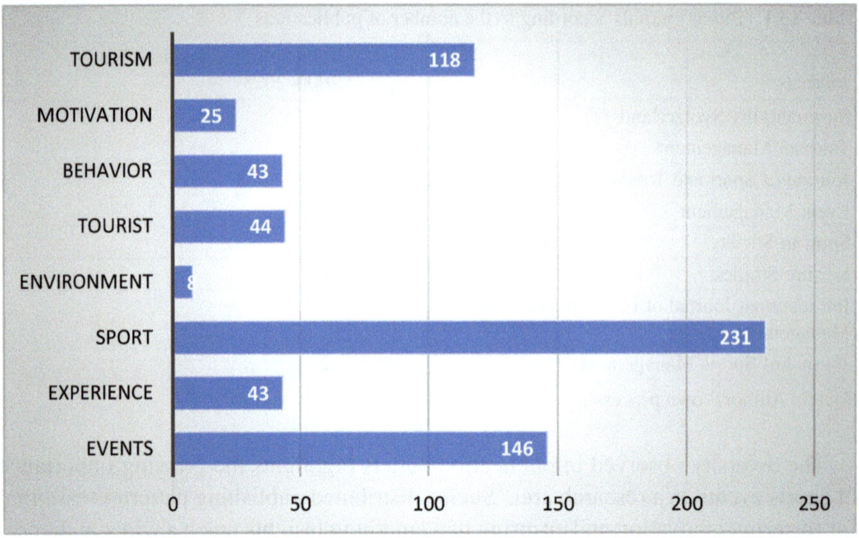

Fig. 13.3 Keywords occurrence in the articles analysed. (Source: Authors' own processing)

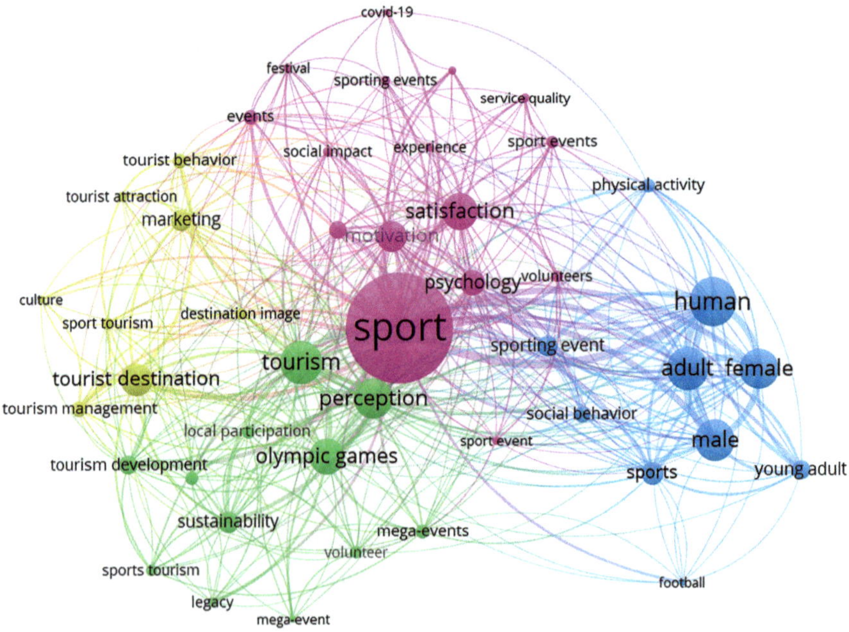

Fig. 13.4 Visualization map—sports events. (Source: Authors' own processing)

evolve within the scientific discourse. By identifying these areas, the analysis contributes to a clearer understanding of the intellectual structure of the field, guiding future research directions and practical applications. The integration of diverse methodologies and thematic areas in this study highlights the multidisciplinary nature of research in sports events, making it a vital resource for academics, industry practitioners, and policymakers alike (Table 13.2).

Cluster 1 reveals that satisfaction emerges as an essential theme, intricately linked to service quality and the vital contributions of volunteer involvement. Motivational and psychological aspects emphasize the influence of intrinsic factors, such as enjoyment and passion, alongside extrinsic drivers like social bonding and rewards, in shaping the overall experience. Moreover, the relationship with social behaviour highlights the significant interplay between group dynamics and personal experiences, demonstrating how collective and individual interactions converge during sports events.

Cluster 2 highlights the significant co-occurrence of the term "sustainability," reflecting a growing emphasis within academia and the sports industry on the long-term impacts and responsible management of events. This underlines the evolving discourse on how sports events can leave enduring benefits, not only through economic gains but also by fostering social cohesion, environmental consciousness, and community development. The research demonstrates a clear shift from viewing sports events as one-off occurrences to recognizing their potential as instruments of sustainable growth.

This cluster suggests an increasing integration of sustainable practices into event planning, addressing critical issues such as resource efficiency, carbon footprint reduction, and community involvement. The analysis highlights how host locations leverage sports events to achieve strategic objectives, such as bolstering tourism flows, driving infrastructure upgrades, and stimulating local economies, while aligning with sustainability goals.

Table 13.2 Cluster's key terms and focus

Cluster (colour)	Key terms	Focus
Cluster 1 (Magenta): Customer Experience and Satisfaction in Sports Events	Satisfaction, experience, motivation, psychology, perception, service quality	This cluster delves into the **emotional and psychological** drivers of customer experience.
Cluster 2 (Green): Mega-Events and Sustainability	Tourism, tourist destination, sustainability, ecotourism, sports tourism	This cluster focuses on the broader ***socio-economic and environmental impacts*** of sports events.
Cluster 3 (Yellow): *Destination Image and Marketing in Sports Tourism and Events*	Marketing, festival, events, tourist behaviour, culture	This cluster highlights the importance of ***strategic event design and promotion***.
Cluster 4 (Blue): Demographic and Behavioural Research in Sports Events	Human, male, female, adult, social behaviour	This cluster explores ***audience segmentation and behavioural patterns***.

Source: Authors' own processing

Moreover, the frequent association with terms such as "tourist destination" and "destination image" emphasizes the role of sports events in branding and tourism development. These events act as powerful catalysts for enhancing the visibility and appeal of host destinations, enabling them to attract a wider audience of tourists, investors, and stakeholders. By crafting a positive and memorable destination image, sports events contribute to reshaping perceptions of cities or countries, elevating their status on the global stage.

Cluster 3 highlights the critical role of marketing strategies and event design in shaping consumer perceptions and behaviours within the realm of sports events. Studies in this cluster explore a broad spectrum of elements, including branding initiatives, the integration of cultural experiences, and consumer engagement efforts such as pre- and post-game events, entertainment offerings, and immersive activities that amplify the overall event experience.

The strong connections between this cluster and those related to tourism and psychology emphasize the multidisciplinary nature of marketing strategies. These connections demonstrate how marketing not only enhances CXs but also plays an essential role in promoting the host destination's appeal. This dual focus underlines the strategic importance of aligning marketing efforts with psychological insights and tourism dynamics to foster engagement and attract diverse audiences.

Furthermore, the link between "marketing" and "experience" highlights the significance of creating memorable moments through designed event atmospheres and compelling promotional strategies. By leveraging tools such as storytelling, interactive activities, and social media outreach, event organizers can create environments that resonate deeply with attendees, ensuring heightened satisfaction, emotional connection, and long-term loyalty.

This cluster also reflects the broader implications of marketing in sports events, including its impact on consumer decision-making, brand equity, and the economic sustainability of host locations. It reinforces the idea that well-executed marketing efforts not only enrich individual event experiences but also contribute to the development of a vibrant, competitive, and socially engaging sports event ecosystem.

Cluster 4 examines how diverse demographic groups, including variations in gender, age, and cultural background, experience and engage with sports events. Studies from this cluster examine behavioural patterns, such as social interactions, levels of physical activity, and participation preferences, to reveal audience-specific trends and tailor event designs to meet diverse needs. These insights allow for the development of strategies that enhance engagement and foster a deeper connection between attendees and the event itself.

The importance of demographic terms within this cluster emphasizes the growing interest in targeted audience strategies. By identifying the preferences, behaviours, and expectations of diverse audience segments, organizers can create more inclusive and adaptive sports events. This approach not only promotes equitable access and enjoyment but also addresses the challenges of engaging underrepresented groups.

The findings are strongly linked to key experiential dimensions such as satisfaction and motivation, highlighting the significant role demographic factors play in shaping perceived value and overall customer engagement. For example, intrinsic motivations like personal enjoyment and extrinsic drivers such as social bonding may manifest differently across age or gender groups, providing unique opportunities for personalized marketing and program design.

Overall, this cluster showcases the importance of understanding demographic diversity as a foundation for strategic planning, inclusive innovation, and improved audience experiences in the rapidly evolving domain of sports event management.

A Content-Based Approach

According to the results of the bibliometric analysis, four dominant clusters emerged in the literature, with Cluster 1—*Customer Experience and Satisfaction in Sports Events* standing out as a significant research area. This cluster reflects the growing academic interest in understanding how audience engage with sports events, their motivations, and the factors influencing their overall experience. Given the relevance of this cluster, the following section delves deeper into the key theories, themes, and findings within this research group, ultimately identifying critical gaps in the existing literature.

Traveling and attending sports events require several criteria, including motivation, time allotment, interest, destination image, prior experiences, personal considerations, and financial circumstances (Getz & Page, 2015). Sports events are attended by people for a variety of reasons. The reasons are typically a mix of general ones (socializing, learning something new, escaping from daily routine, and amusement), as well as ones that rely more on the event's location or the event itself. The motivations behind going to sports events are also closely related to personal demographic traits, including gender, age, place of residence, and socioeconomic status, or group affiliation and family, self-esteem, knowledge, social interaction, entertainment, or escape. Sports events are being used more and more by destination marketers to promote the host destination's brand. By facilitating deeper interaction with touristic destinations and a more comprehensive understanding of the social and cultural life of local communities, the sport and tourism component can enhance the travel experience (Weed & Bull, 2009).

Sport spectatorship has transformed considerably over recent decades, evolving from a simple viewing activity to a complex, multidimensional consumer experience. Funk (2008) provides a foundational understanding of consumer behaviour in sport contexts, establishing that sport spectatorship is not merely about watching athletic competition but also encompasses a holistic consumption experience with emotional, social, and psychological dimensions. This perspective frames sport viewership as an experiential product rather than a simple service transaction.

Getz and Page (2015) extend this conceptualization by positioning sporting events within the broader context of event tourism. Their work emphasizes that sport spectator experiences must be understood within their temporal and spatial dimensions, recognizing that the experience begins before and continues after the actual sporting event. This "pre-during-post" experience framework has significant implications for how sport managers conceptualize service delivery.

The economy has entered a new period that is focused on experiences (Sundbo & Sørensen, 2013), and as a result, tourism research has shifted to focus on the idea of experience (Sfandla & Björk, 2013). Shaw and Ivens (2002) state that 85% of corporate managers support the idea that, in the modern era, CX—rather than price, product, or quality—is the primary competitive advantage. Therefore, understanding the tourist experience should be at the core of both tourism management and sport event tourism management, according to Morgan (2007). Experiences are what modern tourism is all about. For contemporary travellers, a tidy hotel room with a comfortable bed and a satisfying breakfast is no longer sufficient. They need some distinctive and, in some way, special experiences. This characteristic is also applicable to the sports event tourism sector, where visitors can have amazing experiences and adventures by taking part in important sports events. According to Funk (2008), the experience is what drives sport consumer behaviour, whether one is an active (as participant) or a passive (as spectator) attendee in a sports event.

To experience the emotions of incredibly unforgettable sports events, spectators are willing to pay high fees for tickets (Emery et al., 2013). According to Kahle and Riley (2004), group interaction is the primary factor that improves viewers' overall experience during sports events.

The approach of CX varies across different academic disciplines and industry sectors (management, tourism, marketing, e-commerce, sociology, retailing, etc.). Therefore, there is a variation in the definition and comprehension of CX across many contexts and disciplines, even though academics and practitioners generally agree that it plays a critical role in influencing consumer perceptions and behaviours. CX is commonly acknowledged as a comprehensive framework for elucidating and assessing the efficacy of a brand or product offering. It has been acknowledged as a crucial element influencing consumers' future buying behaviour. The term "CX" refers to how consumers feel when interacting with a company's products, services, or environment. Academics and practitioners have acknowledged the importance of CX in generating better value (Nasution et al., 2014; Verhoef et al., 2009).

A review of the CX literature reveals that most CX researchers from different areas use a generalized unimodular perspective across a variety of domains, including tourism management (Ponsignon et al., 2021), service marketing (Bueno et al., 2019), and retailing (Roggeveen & Rosengren, 2022). Over time, numerous CX articles have been published, providing insightful analyses of the CX research phenomenon. Some researchers (e.g., Grewal et al., 2009; Verhoef et al., 2009) examined the role of CX in retailing, others in design and service management (Jain et al., 2017), in experiential marketing or service, in business and management literature, or in hospitality and tourism literature (Kim & So, 2022; Stead et al., 2022). To comprehend its fundamental aspects, researchers have been looking more closely at the CX idea from the customer perspective as it has grown in importance. This approach is also applied to produce pertinent and trustworthy results about the CX concept in the context of sports events.

CX was examined by Greenwell et al. (2002) as a supplementary factor to customer satisfaction at a sports event. The authors underlined that CX represents a culmination of the experiences gained from the main product (such as the game or competition, teamwork), the surroundings and ambiance of the facilities, and the staff at the locations. Other well-known research on CX in sports focuses on managerial perspectives and examines several aspects that impact CX, including team identification and performance, functional components (including technologies), facilities, and service quality, or considers the dynamic nature of the experiences lived (Glebova & Desborders, 2020; Gwinner & Swanson, 2003; Yoshida & James, 2010; Tsuji et al., 2007). Nowadays, sports managers consider the commercialization of such a commodity as sport to be a top priority (Clemes et al., 2011), since fans are more likely to want to return to an event if they are pleased with the level of service provided at the location, the ambiance of the event, or the sport performance (Chanavat & Bodet, 2014). Sports events are viewed by spectators as a journey with several phases (pre-event, event, and post-event) and interactions at each of them. Stadium atmosphere, facility conditions, service quality, language diversity, accessibility, technology, and sharing moments with friends or external factors (e.g., Internet connection, parking slots, rivalry, players' motivation to play hard, and weather) enhance customers' experience.

13.4 Conclusions

The current study implementation is predicated on a theoretical framework for CX that was developed specifically for sports events. Its main goal is to investigate new themes and trends in the study of CETB in relation to sports events. Since Scopus is a globally recognized citation database that provides access to a vast repository of scholarly work that is highly regarded for its dependability, comprehensive scope, and quality of publications, it was chosen as the main source for bibliometric analysis.

Within a set of parameters, it examines the international literature on sports events that was published between 2000 and 2025. With an average below 20 publications per year between 2019 and 2024, scholarly interest in this field increased significantly, establishing its significance as a major area of study. Starting in 2014, interest in these subjects began to sharply increase, emphasizing how popular sports events are becoming in the literature. But, instead of a strong concentration within a small number of journals, the present study reveals a notable dispersion of research across multiple outlets, highlighting that academic discourse on sports events is not dominated by a single journal or a small group of journals.

The study reveals that satisfaction in sports events is driven by service quality, volunteer contributions, and a mix of intrinsic and extrinsic motivations, all shaped by group dynamics and personal interactions. Furthermore, there has been a noticeable change in perspective from considering sports events as isolated incidents to appreciating their potential as tools for long-term sustainable growth. Moreover, the present analysis emphasizes how strategically crucial it is to match marketing initiatives with psychological insights and tourism dynamics. It supports the idea that effective marketing enhances personal event experiences while fostering a dynamic, competitive, and socially connected sports event environment. Another main finding of the research is related to the significance of comprehending demographic diversity as a basis for inclusive innovation, strategic planning, and enhanced audience experiences.

The thematic mapping of the research topic and the examined literature reveals a few important research gaps that require more study and can be classified into six thematic areas:

- *Demographic diversity and segmentation*: limited understanding of how age, gender, and socioeconomic factors influence audience priorities and perceptions of sports experiences;
- *Sustainability integration*: lack of research on how eco-friendly initiatives impact attendee satisfaction and can be seamlessly integrated without compromising event quality;
- *Cross-cultural variances*: limited understanding of how cultural contexts influence CXs and whether experiential drivers are universal or culturally specific;
- *Social impact measurement*: absence of reliable frameworks for measuring sports events' contribution to social cohesion and community involvement;
- *Tourism-sports experience integration*: lack of theoretical models connecting tourism behaviour with sports consumption patterns and understanding broader guest engagement;
- *Technology-human interface balance*: need to assess optimal balance between technological enhancements and authentic human interaction in sports events.

These gaps show how the CX at sporting events has evolved and how more theoretical and empirical research is required.

Differentiation by demographics is one significant divide. Research on the relationship between sport and demographic factors (male, female, and young adult) has been acknowledged, but little is known about how different audience groups prioritize and see various aspects of sports experiences. Knowing how age, gender, and socioeconomic position affect consumer expectations could improve marketing and event organising tactics.

Sustainability in sports experiences is another topic that will get increased attention. Research on how eco-friendly initiatives affect attendee pleasure is lacking, despite the growing recognition of sustainability as a crucial component of event management. Future research should examine the smooth integration of sustainable practices into sporting event experiences without sacrificing essential components of the event. Furthermore, there are currently no reliable frameworks for measuring the social impact of sports events. There is little research measuring the benefits of sports events in promoting social cohesion and community involvement, despite the fact that they are mentioned in some studies. To evaluate how participant experiences contribute to long-term community well-being and transfer into more general social outcomes, a more systematic approach is needed.

Another area that requires more research is cross-cultural variances. Most of the current research focused on regional sports events; however, it is essential to comprehend how CXs vary depending on the cultural setting. Event planners aiming to reach a global audience would benefit greatly from examining whether the fundamental experiential drivers are culturally particular or universal. Furthermore, theoretical models that integrate tourist and sports experiences are still lacking, despite the obvious relationship between the two. The understanding of how sports events influence guest engagement beyond the immediate sporting show may be improved by more thorough frameworks that connect tourism behaviour with sports consumption patterns.

Finally, the role of technology–human interface requires deeper examination. While technological advancements such as digital ticketing, augmented reality, and virtual fan experiences are reshaping sports event engagement, research must assess the optimal balance between technological enhancements and authentic human interaction. Identifying how technology influences emotional and social engagement will be vital for creating enriching sports event experiences in the future.

Addressing these gaps through further research will not only expand academic discourse but also provide practical insights for event planners, tourism stakeholders, and policymakers striving to enhance customer satisfaction in sports events. As sports tourism continues to grow, refining spectator experiences through targeted research will be essential for maintaining competitiveness and ensuring the long-term success of sports events worldwide.

Since this study is one of the few bibliometric studies that focus exclusively on sports events and offer a critical viewpoint on bibliometric research related to sports events in the global literature, it is deemed pertinent to the study's issue.

As limitations of the research, it is relevant to underline that the outcomes of bibliometric analysis often concentrate on measures pertaining to publications, sources, and authors due to methodological constraints. Articles examined in this study are exclusively about sports events, and the fact that their examination is linked only with the ones published in foreign journals available in English language through the Scopus database is another drawback. Since the study relies exclusively on the Scopus database, which creates a single-database bias, this approach may miss relevant studies indexed in other major databases like Web of Science, Google Scholar, or discipline-specific databases, potentially leading to incomplete coverage of the research landscape. The limited keyword scope, "SE = (Sports, Sport, Event, Events)," later refined with "Customer Experience and Tourist Behaviour," may exclude relevant studies that use different terminology or conceptual frameworks to describe similar phenomena. The English-only filter creates a significant limitation by excluding potentially valuable research published in other languages, which could provide important cross-cultural perspectives on sports events, CX, and tourist behaviour.

References

Blomkvist, J., & Segelström, F. (2014). Benefits of external representations in service design: A distributed cognition perspective. *The Design Journal, 17*(3), 331–346.

Bolton, R. N., McColl-Kennedy, J. R., Cheung, L., Gallan, A., Orsingher, C., Witell, L., & Zaki, M. (2018). Customer experience challenges: Bringing together digital, physical and social realms. *Journal of Service Management, 29*(5), 776–808.

Broadus, R. N. (1987). Toward a definition of "bibliometrics". *Scientometrics, 12*, 373–379. https://doi.org/10.1007/BF02016680

Bueno, E., Weber, T. B. B., Bomfim, E. L., & Kato, H. T. (2019). Measuring customer experience in service: A systematic review. *The Service Industries Journal, 39*(11–12), 779–798.

Chalip, L., & Leyns, A. (2002). Local business leveraging of a sport event: Managing an event for economic benefit. *Journal of Sport Management, 16*(2), 132–158.

Chanavat, N., & Bodet, G. (2014). Experiential marketing in sport spectatorship services: A customer perspective. *European Sport Management Quarterly, 14*(4), 323–344.

Chepngetich, B., Ouma, O. K., & Aila, F. O. (2019). Influence of customer experience dimensions on purchase behaviour in Kenyan hotels. *European Journal of Business and Management, 11*(21), 50–62.

Clemes, M. D., Brush, G. J., & Collins, M. J. (2011). Analysing the professional sport experience: A hierarchical approach. *Sport Management Review, 14*(4), 370–388.

De Keyser, A., Lemon, K. N., Klaus, P., & Keiningham, T. L. (2015). A framework for understanding and managing the customer experience. *Marketing Science Institute working paper series, 85*(1), 15–121.

Emery, P. R., Kerr, A. K., & Crabtree, R. M. (2013). Critical incidents, emotions and moments: The London 2012 Olympic spectator experience. In *Proceedings of the 21st conference of the European Association for Sport Management (EASM). Istanbul, 11–15 September 2013.*

Fourie, J., & Santana-Gallego, M. (2011). The impact of mega-sport events on tourist arrivals. *Tourism Management, 32*(6), 1364–1370.

Funk, C. D. (2008). *Consumer behaviour in sport and events*. Elsevier.

Funk, D. C., & James, J. (2001). The psychological continuum model: A conceptual framework for understanding an individual's psychological connection to sport. *Sport Management Review, 4*(2), 119–150. https://doi.org/10.1016/S1441-3523(01)70072-1

Getz, D., & Page, S. J. (2015). Progress and prospects for event tourism research. *Tourism Management, 52*, 593–631.

Glebova, E., & Desborders, M. (2020). Technology enhanced sports spectators customer experiences: Measuring and identifying impact of mobile applications on sports spectators customer experiences. *Athens Journal of Sports, 7*(2), 115–140.

Greenwell, T. C., Fink, J. S., & Pastore, D. L. (2002). Assessing the influence of the physical sports facility on customer satisfaction within the context of the service experience. *Sport Management Review, 5*(2), 129–148.

Grewal, D., Levy, M., & Kumar, V. (2009). Customer experience management in retailing: An organizing framework. *Journal of Retailing, 85*(1), 1–14.

Gwinner, K., & Swanson, S. R. (2003). A model of fan identification: Antecedents and sponsorship outcomes. *Journal of Services Marketing, 17*(3), 275–294.

Homburg, C., Jozić, D., & Kuehnl, C. (2017). Customer experience management: Toward implementing an evolving marketing concept. *Journal of the Academy of Marketing Science, 45*(3), 377–401.

Jain, R., Aagja, J., & Bagdare, S. (2017). Customer experience–a review and research agenda. *Journal of Service Theory and Practice, 27*(3), 642–662.

Kahle, L. R., & Riley, C. (2004). *Sports marketing and the psychology of marketing communication.* Lawrence Erlbaum Associates Inc.

Kaplanidou, K., & Vogt, C. (2007). The interrelationship between sport event and destination image and sport tourists' behaviours. *Journal of Sport and Tourism, 12*(3–4), 183–206. https://doi.org/10.1080/14775080701736932

Kaplanidou, K., & Vogt, C. (2010). The meaning and measurement of a sport event experience among active sport tourists. *Journal of Sport Management, 24*(5), 544–566. https://doi.org/10.1123/jsm.24.5.544

Kim, H., & So, K. K. F. (2022). Two decades of customer experience research in hospitality and tourism: A bibliometric analysis and thematic content analysis. *International Journal of Hospitality Management, 100*, 103082.

Kolb, B. (2006). *Tourism marketing for cities and towns. Using branding and events to attract tourism*. Butterworth-Heinemann.

Lemon, K., & Verhoef, P. (2016). Understanding customer experience throughout the customer journey. *Journal of Marketing, 80*(6), 69–96.

Liu, D. (2015). The image impact of mega-sporting events perceived by international students and their behaviour intentions. *International Journal of Sports Marketing & Sponsorship, 16*(2), 103–118.

Matheson, V. A. (2006). *Mega-events: The effect of the world's biggest sporting events on local, regional and national economies*. Working Paper 06-10, College of the Holy Cross, Department of Economics Faculty Research Series.

Morgan, M. (2007). We're not the Barmy Army!: Reflections on the sports tourist experience. *International Journal of Tourism Research, 9*(5), 361–372.

Morris, S. A., & Van der Veer Martens, B. (2008). Mapping research specialties. *Annual Review of Information Science and Technology, 42*, 213–295. https://doi.org/10.1002/aris.2008.1440420113

Nasution, R. A., Sembada, A. Y., Miliani, L., Resti, N. D., & Prawono, D. A. (2014). The customer experience framework as baseline for strategy and implementation in services marketing. *Procedia – Social and Behavioral Sciences, 148*, 254–261.

Okubo, Y. (1997). *Bibliometric indicators and analysis of research systems: Methods and examples* (OECD Science, Technology and Industry Working Papers, No. 1997/01). OECD Publishing. https://doi.org/10.1787/208277770603

Pike, A. (2013). *Brands and branding geographies*. Edward Elgar.

Ponsignon, F., Lunardo, R., & Michrafy, M. (2021). Why are international visitors more satisfied with the tourism experience? The role of hedonic value, escapism, and psychic distance. *Journal of Travel Research, 60*(8), 1771–1786.

Rai, S., & Singh, K. (2020). A bibliometric analysis of deep web research during 1997–2019. *Journal of Library and Information Technology, 40*(2), 452–460.

Roggeveen, A. L., & Rosengren, S. (2022). From customer experience to human experience: Uses of systematized and non-systematized knowledge. *Journal of Retailing and Consumer Services, 67*, 102967.

Sfandla, C., & Björk, P. (2013). Tourism experience network: Co-creation of experiences in interactive processes. *International Journal of Tourism Research, 15*(5), 495–506.

Shaw, C., & Ivens, J. (2002). *Building great customer-experiences* (pp. 48–66). Palgrave Macmillan. https://doi.org/10.1057/9780230554719

Shtengelov, I. (2020). *Determinants of customer experience in sport events*. https://urn.fi/URN:NBN:fi-fe2020112492843

Small, H. (1973). Co-citation in the scientific literature: A new measure of the relationship between two documents. *Journal of the American Society for Information Science, 24*, 265–269. https://doi.org/10.1002/asi.4630240406

Stead, S., Wetzels, R., Wetzels, M., Odekerken-Schroder, G., & Mahr, D. (2022). Toward multisensory customer experiences: A cross-disciplinary bibliometric review and future research directions. *Journal of Service Research, 25*, 10946705221079941.

Su, X. W., Li, X., & Kang, Y. X. (2019). A bibliometric analysis of research on intangible cultural heritage using CiteSpace. *SAGE Open, 9*(2), 1–18. https://doi.org/10.1177/2158244019840119

Sundbo, J., & Sørensen, F. (2013). *Handbook on the experience economy*. Edward Elgar.
Theodorakis, N. D. (2014). Customer experience in spectator sports. In J. Kandambully (Ed.), *Customer experience management: Enhancing experience and value through service management* (pp. 205–219). Kendall Hunt Publishers.
Theodorakis, N. D., Tsigilis, N., Wann, D. L., Lianopoulos, G., & Al-Emadi, A. (2016). Sport spectator identification scale: An item response analysis approach. *International Journal of Sport Management, 17*(2), 1–19.
Tsuji, Y., Bennett, G., & Zhang, J. (2007). Consumer satisfaction with an action sports event. *Sport Marketing Quarterly, 16*(4), 199–208.
Verhoef, P. C., Lemon, K. N., Parasuraman, A., Roggeveen, A., Tsiros, M., & Schlesinger, L. A. (2009). Customer experience creation: Determinants, dynamics and management strategies. *Journal of Retailing, 85*(1), 31–41.
Vogt, C., & Andereck, K. (2003). Destination perceptions across a vacation. *Journal of Travel Research, 41*, 348–354.
Wakefield, K. L., & Blodgett, J. G. (1996). The effect of the services cape on customers' behavioral intentions in leisure service settings. *Journal of Services Marketing, 10*(6), 45–61.
Wäsche, H., Dickson, G., & Woll, A. (2013). Quality in regional sports tourism: A network approach to strategic quality management. *Journal of Sport & Tourism, 18*(2), 81–97.
Weed, M., & Bull, C. J. (2009). *Sports tourism: Participants, policy and providers* (2nd ed.). Elsevier Ltd. Butterworth-Heinemann.
Yoshida, M., & James, J. D. (2010). Customer satisfaction with game and service experiences: Antecedents and consequences. *Journal of Sport Management, 24*(3), 338–361. https://www.unwto.org
Zhang, C., Moreira, M. R. A., & Sousa, P. S. A. (2020). A bibliometric view on the use of total quality management in services. *Total Quality Management & Business Excellence, 32*, 1–28.
Zupic, I., & Čater, T. (2015). Bibliometric methods in management and organization. *Organizational Research Methods, 18*(3), 429–472.

Chapter 14
Are Romanian Resorts Smart? A Destination-Based Analysis

Cristina Ioana Balint, Iuliana Tudose-Pop, Georgică Gheorghe, Monica Maria Coroş, and Delia Popescu

Abstract This study explores the extent to which Romanian resorts of national interest can be considered "smart destinations" and examines the impact of smart technologies on key tourism indicators. The added value of this research lies in its comprehensive, destination-based analysis of the 54 officially designated resorts of national interest, providing a quantitative assessment of their digitalization and smart features, a topic of growing relevance for tourism competitiveness and sustainability. The research aims to assess the level of smartness across these resorts and to evaluate its influence on tourist arrivals, overnight stays, and average lengths of stay.

A composite Smart Indicator was developed, comprising 39 components such as website presence, multilingual support, mobile applications, social media activity, digital services for tourists and residents, accessibility for people with disabilities, and smart mobility solutions. For each resort, the presence or absence of these features was systematically coded (1/0), based on official sources. Panel data covering the 2001–2023 timeframe were compiled for key tourism metrics. Statistical analysis was conducted using SPSS and XLSTAT, testing the hypotheses regarding the relationship between the Smart Indicator and tourism performance.

C. I. Balint · M. M. Coroş (✉)
Faculty of Business, Universitatea Babeş-Bolyai, Cluj-Napoca, Romania
e-mail: cristina.balint@ubbcluj.ro; monica.coros@ubbcluj.ro

I. Tudose-Pop
Alexandru Ioan Cuza University, Bucureşti, Romania

G. Gheorghe · D. Popescu
Bucharest University of Economic Studies, Bucharest, Romania
e-mail: georgica.gheorghe@com.ase.ro; delia.popescu@com.ase.ro

A. L. Negrusa, M. M. Coroş (eds.), *Innovation, Sustainability, and Growth in a VUCA Environment*, Springer Proceedings in Business and Economics,
https://doi.org/10.1007/978-3-032-11639-0_14

The results reveal significant disparities in smartness levels among the scrutinized resorts, with coastal destinations generally outperforming mountain and balneary resorts. Statistical findings confirm that higher smartness levels are positively associated with increased tourist arrivals, more overnight stays, and longer average lengths of stay, although the magnitude of these effects varies by resort type. The study concludes that advancing smart features is crucial for enhancing the competitiveness, sustainability, and resilience of Romanian resorts. Policy recommendations include targeted investment in digital infrastructure and a focus on inclusive, sustainable smart solutions tailored to each resort's profile.

Keywords Smart tourism destinations · Romanian resorts of national interest · digitalization · destination competitiveness · tourism indicators

14.1 Introduction

Romania's tourism potential is particularly important. The presence of varied forms of relief determines not only particularly attractive landscapes but also numerous soil and subsoil resources that can be capitalized on for tourism purposes. Many natural resources can also be used for therapeutic purposes, such as mineral springs or sapropelic muds. Romania is considered one of the richest European countries in this regard. The way landforms were born generated a series of natural resources that constitute the raw material used for therapeutic purposes. These resources are employed for treatment, or healthcare and maintenance cures in Romania's spa resorts. Along with the spa destinations, the landscapes related to different forms of relief lead to the development of resorts for leisure and recreational activities, both in the mountain areas, where outdoor hiking and trekking activities and winter sports are also associated, and in the coastal areas, where the summer stays associated with water sports and leisure services attract many tourists.

In Romania, at the national level, according to the Governmental Decision (GD) No. 852 of August 13, 2008 (Romanian Government, 2008), updated via the GD No. 108 of February 16, 2024, 54 resorts of national interest have been designated and function as of the current moment (MEET, 2025), according to the corresponding criteria. The resorts on the Romanian seaside concentrate most of the accommodation units in the country (Jugănaru et al., 2025). Their attractiveness increases relative to the desire to spend as much quality free time as possible, with access to new services, in line with the technical progress made. Adapting to change in response to environmental conditions (Polderman et al., 2020) is an important objective to attract tourists connected to these transformations. Adaptation measures that winter-sport activities operators (particularly in the ski sector) can employ refer to technological measures and business practices. In many cases, the period of the COVID-19 health crisis has driven an acceleration of the process of adaptation to new technologies, with positive effects on the modernization of tourist destinations (Foris et al., 2022). Thus, the adaptation of resorts to smart technology has become an essential aspect, crucial in the management of destinations (Panyadee et al., 2023), the digitization process being also present at the resort level, regardless of the profile, becoming a pressing necessity. Tourists will thus take advantage of access to modern technology, adapting their behavior and especially the way they will use the information and buy travel products and services (Romanian Government, 2023). Tourists will also interact more dynamically with the location (Panyadee et al., 2023). Within the framework of Romania's National Strategy for Tourism Development 2023–2035 (Romanian Government, 2023), it is envisaged the modernization of tourist information centers through the use of digital technologies and modern tourism services, the provision of information in several foreign languages, the creation of multimodal urban mobility systems (smart ticketing kiosks, electronic tickets, and information boards), applications, and digital maps of the destinations to help tourists.

The main objective of this chapter is to identify the degree of smartness at the level of the resorts of national interest in Romania. This is the first paper to address the smartness of the 54 destinations of national interest in Romania. According to the authors' knowledge, this study is the first that also clarifies the list of localities included in the categories for which the Romanian National Institute of Statistics collects data (seaside, spa, and mountain destinations). To this end, several smart components have been identified for each resort out of the 54 currently accredited resorts of national interest, by the means of thorough research of the information provided on the official websites of these resorts (MEET, 2025a). The smart dimensions considered were inspired by the six dimensions included in the smart kit offered by the Romanian Smart City Association (ARSC, 2025): smart governance, smart environment, smart mobility, smart economy, smart citizen, and smart living (tourism, culture, and spare time, followed by public spaces, medical services, technology, well-being, and social inclusion). To cover these six dimensions, data were collected from the official websites of the 54 resort destinations of national interest, comprising 39 items that constitute the Smart Indicator developed by the authors, comprising six dimensions: website and multilingual support, mobile applications for tourists and for the local community, social media presence, digital services for residents and tourists, accessibility features for persons with disabilities, smart mobility (e-ticketing and info panels), and environmental sustainability initiatives.

To compute the level of smartness, quantitative research was carried out with the help of the SPSS statistical program. Thus, several correlations were processed to answer the following research questions:

1. Is there a connection between the smart level of the resort and the number of arrivals?
2. Is there a connection between the smart level of the resort and the overnight stays?
3. Which of the categories of resorts analyzed has a better-represented smart component?
4. Does the smart component of the resort contribute to a sustainable environment, or to the growth of the local economy?

The chapter is organized as follows: the introduction to the potential of Romanian tourism and a brief presentation of the smart components of tourism are followed by the literature review section that presents the latest approaches to the research topic and describes Romania's resorts; next comes the presentation of the research method used, followed by the presentation of the results and discussions based on them and at the end a section of conclusions and recommendations.

14.2 Literature Review

Intelligence has already become an important component in the tourism industry (Panyadee et al., 2023), a transformative driver of various operations within it (Sigala, 2018) through artificial intelligence (AI), augmented reality (AR), virtual reality, Internet of Things (IoT), drones, machine learning, robots, and the integration of smart technologies could improve the safety and experience of tourists (Govekar et al., 2025). Future technologies are interconnected through intelligence, data, and connectivity; thus, tourism resources could be converted into smart tourism resources (Sigala, 2018). Smart hotels integrate these advanced technologies (AI, IoT, big data management, and cloud computing) to not only increase productivity and reduce labor costs (Liu et al., 2024) but also to offer new customer experiences. According to Kim et al. (2020), smart hotels are considered an innovative business model within the hotel industry, which, by implementing advanced technologies, makes a difference among competitors. Thus, tourists can easily use the smart hotel's facilities and control different options offered by it, representing a superior option for customers, which has led to an increase in room bookings, with a positive impact on the business (Akel & Noyan, 2024). The use of technology at the level of a smart hotel has the role of increasing efficiency and the tourist experience, thus also contributing to sustainable development.

One of the global problems facing humanity is climate change. Tourism is an activity that, to a large extent, contributes, and fully feels the effects of climate change, both at the level of mountain resorts and in seaside or spa resorts. A factor that contributes to the accentuation of this phenomenon is also the greenhouse gas emissions, which are related to tourism activity, both globally and locally. Although mountain or seaside tourism presents a seasonal activity at the discretion of Romania, the peak seasons of the season, both summer and winter, are accompanied by intense road traffic, which affects the environment through the gas emissions produced, or through high energy consumption in resorts, which resembles crowded cities. In these situations, tourist resorts can be regarded and approached like smart cities, especially in terms of efficient and sustainable energy management, through innovation and smart strategies' implementation (Polderman et al., 2020). Digital technologies also help tourism destinations manage resources more efficiently, support sustainable development, and increase the quality of life for both tourists and residents (Homeghi et al., 2024). The increase in seawater temperature, as a result of climate change, leads to a degradation of the marine ecosystem or to a reduction of the beach area (Dumitrescu et al., 2021). Seaside resorts must adapt and modernize their infrastructure, offering tourists new opportunities to spend their holidays safely, thus attracting a greater number of tourists (Popescu et al., 2020).

The role of ICT during the COVID-19 pandemic has been fundamental in terms of access to services, for both residents and tourists (Cruz-Estrada et al., 2022) within tourism destinations. Tache et al. (2022) present a GIS IT application, which provides digital information services, useful for travelers who choose multimodal transport (rail and road-based solutions) to tourism resorts in Romania.

Mountain tourism resort has a mountain ambience, which includes, in addition to natural environmental factors, accommodation structures suitable for the specific function of tourism, rest, and sport (Erdeli et al., 1999). In the case of winter sports resorts, there are amenities and arrangements that correspond to the practice of winter sports. Romania has particularly important natural resources (Bacoş & Gabor, 2020), and the Carpathian chain, which occupies approximately 30% of the country's surface, represents an advantage in terms of winter sports (Ciangă & Răcăşan, 2015). Thus, the first arrangements in the mountain area for tourism appeared at the end of the second century, in Sinaia, then in Buşteni, Predeal, and Cheia. Subsequently, their number increases by diversifying the offer for different activities, such as relaxation, mountain hiking, and winter sports in Poiana Braşov, Borşa, Voineasa, or Azuga (Bacoş & Gabor, 2021). The rapid growth of digital technologies has led to an entire transformation process in mountain resorts that are adopting more and more digital solutions to improve operational efficiency and reduce disparities among the resorts in Romania and those in other countries (Bidault, 2020; Homeghi et al., 2024).

Winter sports and how the ski area is affected are topics addressed by Herman et al. (2021), in the context of climate change, as well as their contribution to the local economy. The research results, reported to the resorts of national interest found in the analysis, show a contribution to the local economy of less than 5% in the resorts of Borsec, Câmpulung Moldovenesc, Azuga, Voineasa, and Poiana Braşov, between 5% and 10% for Vatra Dornei, 10–20% for Sinaia and Busteni, and 20–30% for Slănic Moldova and Predeal (Herman et al., 2021). Thus, Oprea and Briciu (2023) present, from the point of view of climate change, Vatra Dornei and Câmpulung Moldovenesc resorts, with the longest ski slopes, considered to be the most important in Suceava County. Romania has 44 ski resorts (Herman et al., 2021; Bacoş & Gabor, 2020). However, Romania is an undervalued winter sports destination, comprising today a less than 210 km (MEET, 2025b).

Medical tourism, driven by smart health systems, has recently seen an impressive growth (Dar & Kashyap, 2022). Treatments applied to medical tourists have been facilitated by smart technologies, and in the COVID-19 pandemic, this has been fully acknowledged. As a result, smart healthcare systems and the need for innovation in healthcare are becoming increasingly indispensable. Spa tourism, which, in the past, focused only on health tourism and rehabilitation, must adapt to the challenges it faces and must adapt its offers and present new tourism concepts according to Arsi'c et al. (2024). Spa resorts, which have an old worldwide tradition, and also in Romania, have several components, such as spa, health, and wellness. In Romania, there are more than 160 localities that have therapeutic mineral resources (World Bank, 2019). Although the spa component has been documented since Roman times, today, together with the wellness component, they constitute forms of modern tourism that ought to support sustainable development and be integrated into the natural environment. The concept of wellness can be extended to all tourist products that involve a stay, thus completing the tourist's daily balance.

In the case of seaside tourism, the main components are both natural and infrastructure, targeting beaches, accommodation, and leisure units. Seaside tourism has grown a lot in terms of demand, constituting a challenge for the management of the destination; at the same time, it also constitutes an opportunity to improve services, in general, and also those provided on the seashore or by the beaches. Safety and security in the tourism industry are critical concerns for seaside tourism (Govekar et al., 2025). Beaches, in order to increase the quality of the tourist experience, offer opportunities to implement technological solutions to streamline the decision-making process (Cruz-Estrada et al., 2022).

In their bibliometric analysis, El Archi and Benbba (2023) highlighted the research directions related to the adoption of digital technology in sustainable tourism destinations and pointed toward the complexity of this field, concluding that various stakeholders, including tourism operators, policymakers, local communities, and technology providers, must cooperate for the development of the destination. This is perfectly valid for the case of the Romanian tourist resorts of national interest. The same paper identifies varied areas of study, like the use of digital technology in tourism marketing and management, the environmental impacts of digital technology, and the enhancement of the overall tourist experience. Furthermore, the same authors also mention the need to integrate digital technology in tourism planning, destination management, and sustainable development (El Archi et al., 2023a).

14.3 Research Methodology

Previous research aimed at smart resorts or specific components that are integrated into the resort (hotels, beaches, ski area, and other types of tourism-specific infrastructure) was carried out using both quantitative methods (most often questionnaires were used as a data collection tool) or qualitative methods (semi-structured interviews with stakeholders and staff, staff engaged in tourist activities or with tourists). Furthermore, the analysis of smart resorts has been amplified, especially during the COVID-19 pandemic and after it, when the usefulness and need for permanent adaptation to new technologies increased significantly.

Cruz-Estrada et al. (2022) conducted quantitative research, using multiple regression, which highlights the fact that the tourists' perception of confidence in visiting the beaches of the municipality where the research was conducted is related to variables such as technologies related to social distancing (during the COVID-19 pandemic), beach access control, or other information regarding the tourist area. Volkmann et al. (2021) conducted qualitative research through semi-structured interviews, applied in hotels in the resorts of Sinaia, Buşteni, Azuga, and Slănic Prahova, to examine the actions of tourism stakeholders to attract foreign tourists during the COVID-19 pandemic. Two resorts, namely, Sinaia and Predeal, that are among the best known in Romania, along with other examples, were the subject of research by Homeghi et al. (2024). The authors analyzed 13 features related to digital technology, also comparing the resorts in Romania with resorts in other countries targeted in the research. Following the research, Sinaia resort turned out to have a stronger digital integration compared to the Predeal resort.

Employing a systematic literature review study, El Archi and Benbba (2023) identified the contexts of the Technology Acceptance Model and of the Unified Theory of Acceptance and Use of Technology in tourism and hospitality research, namely: mobile applications, social media marketing, AR, food and beverage, tourism experience, mapping, and others. Another study identified emerging research areas, such as the achievement of sustainable development goals, the adoption of green IT, smart energy, and waste management (El Archi et al., 2023b).

In this chapter, the research aimed at identifying the implications that smart technologies have on the main tourism indicators, such as the number of arrivals, the number of overnight stays, and the average length of stay at the destination level.

In the first stage, a Smart Indicator was created consisting of 39 components (most of them closely related to the Smart Destination concept, as previously identified and discussed by Coroş et al. (2023, 2024) and inspired by the studies of El Archi and Benbba (2023) and of El Archi et al. (2023a, b)). Among the 39 components, several can be mentioned here: the presence of a website, the foreign languages available for the website, smart applications for the tourist resort, the presence on the main social media networks, the digital services offered to the inhabitants, the existence of a destination brand, the types of strategies developed and implemented, the managerial approaches to heritage valorization, the services available for people with disabilities, and many others. For each of the 54 resorts of national interest in Romania, the presence or absence of the 39 components was verified and quantified. If a component was present, it was denoted with 1; otherwise, it was denoted with 0. In the end, all the identified values were summed up, and thus, the Smart indicator was created for all 54 destinations. The Smart Indicator aggregates 39 items under the six smart dimensions (website and multilingual support, mobile applications for tourists and for the local community, social media presence, digital services for residents and tourists, accessibility features for persons with disabilities, smart mobility (e-ticketing and info panels), and environmental sustainability initiatives.

Table 14.1 shows the results of the Smart Indicator created.

Regarding the quantification of the tourism indicators that were of interest for the current research, the values from 2001 to 2023 were taken into consideration. The last year was 2023 because at the time when the research was conducted, not all the necessary data for 2024 were available.

To highlight the influence of the Smart Indicator on the tourist indicators of interest, all the data were organized on the "data panel" system. This way of organizing

Table 14.1 Smart Indicator for the 54 resorts of national interest in Romania

Number	County	Town	Smart indicator (total points)
1	Arad	Moneasa	8
2	Argeş	Dâmbovicioara	13
3	Bacău	Slănic-Moldova	12
4	Bacău	Târgu Ocna	20
5	Bihor	"Zona turistică Centrul Istoric—Coridorul Crişul Repede, municipiul Oradea"	26
6	Bihor	Băile Felix	15
7	Bistriţa-Năsăud	Sângeorz-Băi	9
8	Brăila	Lacu Sărat	3
9	Braşov	Poiana Braşov	25
10	Braşov	Predeal	13
11	Braşov	Râşnov	15
12	Caraş-Severin	Zona Borlova-Muntele Mic, comuna Turnu Ruieni	9
13	Caraş-Severin	Băile Herculane	14
14	Constanţa	Costineşti	13
15	Constanţa	Mamaia	25
16	Constanţa	Mangalia	22
17	Constanţa	Cap Aurora	17
18	Constanţa	Jupiter	18
19	Constanţa	Saturn	18
20	Constanţa	Venus	16
21	Constanţa	Neptun-Olimp	14
22	Constanţa	Eforie Nord	24
23	Constanţa	Eforie Sud	23
24	Constanţa	Zona Mamaia Nord, oraşul Năvodari	28
25	Constanţa	Techirghiol	14
26	Covasna	Covasna	23
27	Dâmboviţa	Zona Peştera-Padina, comuna Moroeni	17
28	Dâmboviţa	Pucioasa	16
29	Harghita	Băile Tuşnad	25
30	Harghita	Borsec	18
31	Hunedoara	Zona Parâng-Petroşani	18
32	Hunedoara	Geoagiu-Băi	19
33	Ialomiţa	Amara	12
34	Maramureş	Ocna Şugatag	16
35	Maramureş	Borşa	17
36	Maramureş	Vişeul de Sus	21
37	Mureş	Sovata	19

(continued)

Table 14.1 (continued)

Number	County	Town	Smart indicator (total points)
38	Neamţ	Zona turistică a municipiului Piatra-Neamţ	26
39	Neamţ	Târgu-Neamţ	25
40	Prahova	Azuga	18
41	Prahova	Busteni	15
42	Prahova	Sinaia	21
43	Prahova	Slănic	8
44	Satu Mare	Tăşnad	14
45	Suceava	Suceviţa	18
46	Suceava	Câmpulung Moldovenesc	26
47	Suceava	Vatra Dornei	25
48	Suceava	Gura Humorului	14
49	Timiş	Buziaş	16
50	Vâlcea	Voineasa	8
51	Vâlcea	Băile Govora	18
52	Vâlcea	Băile Olăneşti	15
53	Vâlcea	Călimăneşti-Căciulata	15
54	Vâlcea	Horezu	16

Source: Authors' processing

data has the capacity to highlight the implications of the Smart Indicator on the indicators under scrutiny.

At the beginning of the research, the objective of the research was stated, namely: to investigate the influence of smart technologies upon the main tourism indicators at the level of resorts of national interest in Romania. To achieve this goal, three research questions were formulated as follows:

RQ1: How does the level of smart development of tourism resorts of national interest influence the number of tourist arrivals throughout the 2001–2023 timeframe?
RQ2: Is there a significant relationship between the Smart Indicator and the total number of overnight stays of tourists, depending on the type of resort?
RQ3: What is the effect of the Smart Indicator upon the average length of stay of tourists in the different types of tourist resorts of national interest?

As established in the literature, for each research question, working hypotheses were formulated and further tested with the help of XLSTAT.

Thus, for the present research, the following working hypotheses were considered:

H1: A higher level of the Smart Indicator is associated with a higher number of tourist arrivals in tourism resorts of national interest.

This hypothesis is supported by previous studies showing that the integration of smart technologies, such as digital information platforms, mobile apps, and online booking services, enhances destination attractiveness and accessibility, leading to increased tourist flows (Sigala, 2018; Panyadee et al., 2023). Empirical evidence from Akel and Noyan (2024) and Kim et al. (2020) further confirms that smart features in hospitality and destinations positively impact visitation rates.

H2: The Smart Indicator positively influences the number of overnight stays of tourists, with significant variations depending on the type of resort.

This is grounded in research indicating that smart solutions (e.g., digital guest services, personalized recommendations, and enhanced safety measures) improve the tourist experience and satisfaction, thus encouraging longer stays (Homeghi et al., 2024; Cruz-Estrada et al., 2022). The effect may differ between mountain, seaside, and balneary resorts due to their specific tourism profiles and visitor expectations.

H3: There is a positive relationship between the smartness level and the average length of stay of tourists, influenced by the specifics of each type of resort.

The literature suggests that digitalization and smart destination management contribute to a richer and more convenient visitor experience, which can contribute to the extension of the duration of stays (Popescu et al., 2020; Volkmann et al., 2021). Enhanced digital infrastructure and smart services are particularly effective in supporting sustainable tourism development and increasing resort competitiveness (Guvernul României, 2023; Polderman et al., 2020).

14.4 Results and Discussions

For the number of Arrivals, respectively, the number of Overnight stays, the data were collected from the TEMPO Online platform (National Institute of Statistics, 2025), while the indicator "Average length of stay" was calculated by dividing the number of overnight stays by the number of arrivals. Throughout this analysis, a dummy variable was created; this consisted of the resort destinations coded with numerical values, by type of resort. This coding was possible due to the fact that the central office of the National Institute of Statistics (2025) provided the researchers with the nomenclature of localities, which enabled the identification of the exact type of resort under which each of the 54 localities is recorded.

Once the coding process had concluded, hypotheses were further tested.

To test the first hypothesis (H1)—according to which the SMART indicator significantly influences the number of tourist arrivals in tourist resorts of national interest—a panel model with bidirectional random effects (two ways) was estimated. The dependent variable was the annual number of tourist arrivals, and the independent variables included the SMART indicator, the type of resort, and an interaction variable between them.

The results obtained ($Chi^2 = 221.62$, $p < 0.0001$) (Table 14.2) indicate that at least one of the explanatory variables has a significant effect on the number of arrivals. The coefficient of determination ($R^2 = 0.168$) suggests that approximately 16.8% of the variation in the number of tourist arrivals is explained by the model.

Table 14.2 Results of the panel regression for the dependent variable "Tourist Arrivals"

Variable	Estimated coef.	*p*-value	Interpretation
SMART Indicator	11.620	<0.0001	Positive and significant effect
Resort type	8.911	0.421	Insignificant effect
SMART Indicator × type of resort	−1.292	0.030	Moderately negative, significant effect (negative interaction)
Adjusted R^2: 0.168			
Chi^2 test:	$\chi^2 = 221.620$	<0.0001	

Source: Authors' processing

The analysis of the coefficients reveals that the SMART indicator has a positive and statistically significant effect on tourist arrivals (coefficient = 11.619, $p < 0.0001$). Therefore, an increase of one point in the SMART indicator is associated, on average, with an increase of about 11,620 tourist arrivals per year. This relationship supports the hypothesis that the degree of "intelligence" of a resort (measured by the SMART indicator) contributes to increasing tourist attractiveness.

The type of resort, individually analyzed, does not have a statistically significant effect on arrivals (coefficient = 8.911, $p = 0.421$). However, the interaction between the SMART indicator and the resort type is negative and significant (coef. = −1.292, $p = 0.030$), indicating that the positive effect of the SMART indicator on arrivals is partially moderated by resort type. In other words, the increase in the SMART indicator produces different effects depending on the specifics of the resort (spa, mountain, seaside, etc.), being lower in certain types of destinations.

The results are also in line with those of other studies, such as those carried out by Ji et al. (2024), who showed that designating a city as "smart" led to a 24.3% increase in tourism revenues and a 17.2% increase in tourist arrivals. Similarly, a study conducted in 2024 in Ho Chi Minh City on smart technologies in tourism highlights that tools such as mobile apps and AR increase tourist satisfaction and their intention to return (Trần Thị Xuân et al., 2024).

To test the second hypothesis (H2), according to which there is a positive and significant relationship between the SMART indicator and the number of tourist overnight stays, a panel model with bidirectional random effects (two-way) was estimated. The model included, as a dependent variable, the number of overnight stays, and as explanatory variables: the SMART indicator, the type of resort, and the interaction between them.

The overall test of statistical significance indicates that the model is significant ($\text{Chi}^2 = 134.23$, $p < 0.0001$) (Table 14.3), and the coefficient of determination $R^2 = 0.109$ shows that approximately 10.9% of the variation in overnight stays is explained by the independent variables included in the model.

The results show that the SMART indicator positively and significantly influences the number of overnight stays (coef. = 37.924, $p < 0.0001$). This means that a one-point increase in the SMART indicator is associated, on average, with an increase of about 37,924 overnight stays annually in tourist resorts. This

Table 14.3 The results of the panel regression for the dependent variable "Number of Tourist Overnight Stays"

Variable	Estimated Coef.	*p*-Value	Interpretation
SMART Indicator	+37.924	<0.0001	Positive and significant effect
Resort type	+70.365	0.072	Marginally significant positive influence
SMART Indicator × type of resort	−7.008	<0.001	Moderately negative, significant effect (negative interaction)
Adjusted R^2: 0.109			
Chi² test:	$\chi^2 = 134.225$	<0.0001	

Source: Authors' processing

Table 14.4 Results of the panel regression for the dependent variable "Average Length of Stay"

Variable	Estimated Coef.	*p*-Value	Interpretation
SMART Indicator	0.209	<0.0001	Positive and significant effect
Resort type	2.998	<0.0001	Positive and significant effect
SMART score × type of resort	−0.153	<0.0001	Moderately negative, significant effect (negative interaction)
Adjusted R^2: 0.097			
Chi² test:	$\chi^2 = 117.499$	<0.0001	

Source: Authors' processing

relationship confirms the H2 hypothesis, emphasizing the importance of integrating smart solutions in the development of destinations.

The term of interaction between the SMART score and the type of resort is negative and significant (coefficient = −7.008, $p = 0.001$), indicating that the positive effect of the SMART score on overnight stays is diminished in certain types of resorts. Therefore, the efficiency of investments in smart infrastructure can be different depending on the specifics of each tourist destination.

The results of our study are supported by a series of recent studies that highlight the impact of smart infrastructure on tourism performance. For example, the analysis of Benidorm's transformation into a smart tourist destination shows a positive correlation between investments in smart infrastructure and improved occupancy rate and resource efficiency (Aguirre et al., 2022). Other research on regional tourism competitiveness highlights the importance of smart specialization (Romão, 2021).

To test the third hypothesis (H3)—according to which the SMART score significantly influences the average length of stay in tourist resorts of national interest—a panel model with bidirectional random effects (two-way) was applied. The dependent variable was the average length of stay, and the predictors included were the SMART score, the type of resort, and the interaction between the two. The model is statistically significant (Chi² = 117.499, $p < 0.0001$) (Table 14.4), and the coefficient of determination $R^2 = 0.097$ suggests that approximately 9.7% of the variation in average length of stay is explained by the variables included in the model. Although the R^2 value is relatively modest, it is frequently found in panel analyses in the social sciences, where tourism behaviors are influenced by a multitude of factors that are difficult to fully capture in an econometric model.

The results highlight that the SMART score has a positive and significant effect on the average length of stay (coef. = 0.209, $p < 0.0001$), which means that an increase of one point in the SMART score is associated with an average increase of 0.209 days in the length of stay of tourists in the resort. This association supports the H3 hypothesis, suggesting that the integration of smart solutions can help extend the stay in tourist destinations. Also, the type of resort has a positive and significant effect (coefficient = 2.998, $p < 0.0001$), indicating that the average length of stay differs significantly depending on the specifics of the resort (spa, mountain, coast, etc.), with some types of resorts favoring longer stays.

In contrast, the interaction between the SMART score and the type of resort has a negative and significant effect (coef. = −0.153, $p < 0.0001$). It suggests that the positive influence of the SMART score on length of stay decreases in certain types of resorts, probably due to specific functional or seasonal characteristics (e.g., in seaside resorts, digitalization can increase comfort, but not the length of vacation, which is influenced by climatic and calendar factors).

The findings of the paper are also supported by the conclusions of previous studies in the literature. Thus, Oklevik et al. (2021) show that cultural attributes and wellness facilities, frequently improved by smart technologies, positively influence the length of tourists' stay, confirming the link between the digitized tourist offer and the extension of the stay. In the same direction, Shen et al. (2020) emphasize that Smart technologies contribute significantly to improving the tourist experience both in the active and planning phases of the visit, which leads to a higher level of engagement and, implicitly, longer stays.

The results of the panel analyses highlight a significant influence of the SMART score on the main indicators of tourist demand in tourist resorts of national interest in Romania. Therefore, for H1, the SMART score has a significant positive impact on the number of tourist arrivals, although the effect is moderate and influenced by the type of resort. In the case of H2, the SMART score positively and significantly influences the number of overnight stays of tourists, reinforcing the hypothesis that a higher level of tourist "intelligence" attracts a higher volume of tourists who choose to stay overnight. As for H3, the SMART score has a direct positive effect on the average length of stay, but the negative interaction with the type of resort suggests that this effect varies depending on the profile of the destination—being lower in spa and mountain resorts than in seaside or mixed resorts.

Therefore, the data support the idea that the integration of Smart components in a tourist destination is associated with a quantitative increase in tourist demand, but the effects vary depending on the specifics of the resort, which suggests the need for development policies adapted to the local profile.

14.5 Conclusions and Recommendations

The present study offers a comprehensive destination-based analysis of the integration of smart technologies in the 54 officially designated resorts of national interest in Romania. By developing a composite Smart Indicator encompassing 39 distinct digital and smart features, ranging from online presence and multilingual support to advanced digital services and smart mobility solutions, the research provides a robust quantitative assessment of destination smartness and its influence on key tourism indicators.

The findings reveal significant disparities in the levels of smartness among Romanian resorts. Coastal destinations generally outperform their mountain and beach counterparts, reflecting a more advanced adoption of digitalization and smart technologies. This is likely due to the higher concentration of accommodation units and the increased exposure to international tourism trends along the Black Sea coast. Statistical analyses confirm that higher values of the Smart Indicator are positively associated with increased tourist arrivals, more overnight stays, and longer average lengths of stay, although the magnitude and significance of these effects vary by resort type. These results underscore the importance of digital transformation to enhance the attractiveness and competitiveness of tourist destinations.

Furthermore, the study highlights the critical role of smart technology adoption in supporting sustainable development and resilience within the tourism sector. The COVID-19 pandemic accelerated the need for digital adaptation, which makes smart solutions essential for safe, efficient, and responsive destination management. The integration of smart features not only improves the tourist experience and contributes to more efficient resource management and environmental sustainability, aligning with global best practices and the objectives of the Romanian National Strategy for Tourism Development 2023–2035.

In conclusion, advancing smart features in Romanian resorts is not merely a matter of technological innovation but a strategic imperative to ensure long-term competitiveness, sustainability, and resilience. Targeted investments in digital infrastructure, tailored to the unique profiles of coastal, mountain, and balneary resorts, are essential to maximize the benefits of smart tourism. Policymakers and destination managers must prioritize inclusive and sustainable smart solutions, using stakeholder engagement and continuous monitoring to drive positive outcomes for both tourists and local communities.

In light of the findings presented throughout this chapter, several targeted ***proposals and recommendations*** emerge to promote the advancement of smart tourism in Romanian resorts of national interest. These recommendations are designed to address the identified gaps in digitalization and smart technology adoption, while also supporting the broader objectives of competitiveness, sustainability, and resilience.

The first recommendation focuses on the *prioritization of investments in the digital infrastructure*. Thus, a fundamental step is the allocation of resources to the development and modernization of digital infrastructure tailored to the unique needs of each resort category: coastal, mountain, and balneary. This includes expanding the coverage of high-speed internet, deploying digital tourism information platforms, and ensuring the availability of mobile applications that improve the tourist experience. Such investments should be guided by a clear understanding of the specific challenges and opportunities faced by each destination, ensuring that the benefits of digitalization are maximized across the board.

The second recommendation highlights the importance of *developing and promoting inclusive and sustainable smart solutions* at the resort-destination level. In this respect, smart tourism initiatives must be inclusive, ensuring that all visitors, including those with disabilities, can fully access and benefit from digital services. Multilingual support, accessible websites and apps, and the integration of universal design principles are essential components of this approach. Furthermore, smart solutions must be aligned with sustainability goals, incorporating features such as energy-efficient management systems, digital ticketing for public transport, and real-time monitoring of environmental indicators. These measures not only improve the quality of life for tourists and residents but also contribute to the long-term sustainability of destinations.

A third direction to consider is the *alignment of the destination's policy and strategy with the national development goals*. Therefore, local and national authorities ought to align their policies with the objectives outlined in the National Strategy for Tourism Development 2023–2035. This includes modernizing tourist information centers through digital technologies, developing multimodal urban mobility systems (such as smart ticketing kiosks and electronic tickets), and providing digital maps and applications to tourists. Policy frameworks should also encourage public–private partnerships to leverage expertise and resources for the implementation of smart tourism projects.

The fourth key point to bear in mind is that of *strengthening stakeholder engagement and capacity building*. In this sense, the effective adoption of smart technologies requires the active participation of all relevant stakeholders, including resort operators, local authorities, technology providers, and the local community. Training programs and workshops should be organized to build digital competencies among staff and residents, ensuring that new technologies are used effectively and efficiently. Engaging stakeholders in the planning and implementation process will foster a sense of ownership and support for smart tourism initiatives.

The fifth recommendation emphasizes the need to *develop and implement continuous monitoring and evaluation mechanisms*. Thus, to ensure that smart tourism initiatives deliver the intended benefits, it is essential to establish robust mechanisms for monitoring and evaluation. Data-driven approaches should be employed to evaluate the impact of smart technologies on key tourism indicators, including arrivals, overnight stays, and the average length of stay. Regular reviews will allow policymakers and destination managers to identify best practices, address emerging challenges, and adapt strategies as needed.

By implementing these recommendations, Romanian resorts can improve their attractiveness, improve the quality of services, and position themselves as competitive and sustainable destinations in the evolving global tourism landscape.

References

Aguirre, A., Zayas, A., Gómez-Carmona, D., & Sánchez, J. A. L. (2022). Smart tourism destinations really make sustainable cities: Benidorm as a case study. *International Journal of Tourism Cities, 9*, 51–69. https://doi.org/10.1108/IJTC-01-2022-0006

Akel, G., & Noyan, E. (2024). Exploring the criteria for a green and smart hotel: Insights from hotel managers' perspectives. *Journal of Hospitality and Tourism Insights, 7*(5), 2992–3012. https://doi.org/10.1108/JHTI-08-2023-0555

Arsi'c, M., Vujko, A., & Kneževi'c, M. (2024). Development perspectives on wellness and spa tourism in the context of tourism business sustainability. *Sustainability, 16*, 8760. https://doi.org/10.3390/su16208760

Bacoş, I. B., & Gabor, M. R. (2020). Consumers' preferences of winter tourist packages in Romania: A quantitative case study. *Annals of "Dunărea de Jos" University of Galati Fascicle I. Economics and Applied Informatics, XXVI*(3), 157–164. https://doi.org/10.35219/eai15840409150

Bacoş, I. B., & Gabor, M. R. (2021). Tourism economy. Mountain tourism: Quantitative analysis of winter destinations in Romania. *Economics, 9*, 143–159. https://doi.org/10.2478/eoik-2021-0005

Bidault, Y. (2020). How can ski resorts get smart? Transdisciplinary approaches to sustainable winter tourism in the European Alps. *Sustainability, 12*(14), 5593. https://doi.org/10.3390/su12145593

Ciangă, N., & Răcăşan, B. S. (2015). Ski areas and slopes in Romania. Reviewing current state of winter sports tourism unfolding possibilities within Carpathian Mountains. *Studia UBB Geographia, LX(1)*, 157–173.

Coroş, M. M., Popescu, D., Tudose (Pop), I., & Gheorghe, G. (2023). Smart city, smart tourism. In A. L. Negruşa & M. M. Coroş (Eds.), *Remodelling businesses for sustainable development, 2nd International Conference on Modern Trends in Business, Hospitality, and Tourism, Cluj-Napoca, Romania, 2022* (pp. 167–188). Springer Nature. https://doi.org/10.1007/978-3-031-19656-0_12

Coroş, M. M., Popescu, D., Tudose (Pop), I., & Gheorghe, G. (2024). Are Small Towns Smart Destinations? In A. L. Negruşa & M. M. Coroş (Eds.), *Sustainable approaches and business challenges in times of crisis, 3rd International Conference on Modern Trends in Business, Hospitality, and Tourism, Cluj-Napoca, Romania, May 4–6, 2023* (pp. 203–224). Springer Nature. https://doi.org/10.1007/978-3-031-48288-5_12

Cruz-Estrada, I., Miranda-Zavala, A. M., & Ramírez-Torres, M. (2022). From the perspective of the visitor, smart beach management: Rosarito, Baja California, Mexico. *Smart Tourism, 3*(1), 1722. https://doi.org/10.54517/st.v3i1.1722

Dar, H., & Kashyap, K. (2022). Smart healthcare system (SHS): Medical tourism delivering, consumption, and elevating tool in the ages of smart technologies. *Tourism Planning & Development, 20*(3), 397–415. https://doi.org/10.1080/21568316.2022.2109206

Dumitrescu, G.-C., Poladian, S. M., & Aluculesei, A.-C. (2021). Repositioning of Romanian seaside tourism as an effect of climate change. *Information, 12*, 108. https://doi.org/10.3390/info12030108

El Archi, Y., & Benbba, B. (2023). The applications of technology acceptance models in tourism and hospitalityresearch: A systematic literature review. *Journal of Environmental Management and Tourism, XIV, Spring*(2(66)), 379–391. https://doi.org/10.14505/jemt.v14.2(66).08

El Archi, Y., Benbba, B., Kabil, M., & Dávid, L. D. (2023a). Digital technologies for sustainable tourism destinations: State of the art and research agenda. *Administrative Sciences, 13*, 184. https://doi.org/10.3390/admsci13080184

El Archi, Y., Benbba, B., Nizamatdinova, Z., Issakov, Y., Vargáné, G. I., & Dávid, L. D. (2023b). Systematic literature review analysing smart tourism destinations in context of sustainable development: Current applications and future directions. *Sustainability, 15*(6), 5086. https://doi.org/10.3390/su15065086

Erdeli, G., Cândea, M., Braghină, C., Costachie, S., & Zamfir, D. (1999). *Dicţionar de Geografie umană*. Corint.

Foris, D., Bâra, G. R., & Mălăescu, S. (2022). Tourism professionals' awareness of the new technologies: A premise of the destinations' modernization and development. In J. L. Reis, E. P. López, L. Moutinho, & J. P. M.d. Santos (Eds.), *Marketing and smart technologies. Smart innovation, systems and technologies* (Vol. 279, pp. 527–538). Springer. https://doi.org/10.1007/978-981-16-9268-0_44

Govekar, S., Dayanand, M. S., & Dsouza, E. P. (2025). Optimizing beach tourism through smart solutions: A comprehensive review of smart technologies and sustainable strategies. *Journal of Tourism, 38*, 2803.

Herman, G. V., Grama, V., Buhas, S., Garai, L. D., Caciora, T., Grecu, A., Gruia, A. K., Hudea, O. S., & Peptenatu, D. (2021). The analysis of the ski slopes and the degree of economic dependence induced by winter sports tourism. The case of Romania. *Sustainability, 13*, 13698. https://doi.org/10.3390/su132413698

Homeghi, A. F., Ivasciuc, I. S., & Ispas, A. (2024). Comparative study on the use of digital technologies in mountain resorts: An analytical approach. *Bulletin of the Transilvania University of Braşov. Series V, Economic Sciences, 17(66)*(2). https://doi.org/10.31926/but.es.2024.17.66.2.4

Ji, X., Chen, J., & Zhang, H. (2024). Smart city construction empowers tourism: Mechanism analysis and spatial spillover effects. *Humanities and Social Sciences Communications, 11*, 1–14. https://doi.org/10.1057/s41599-024-03626-w

Jugănaru, M., Jugănaru, I. D., Moraru, A.-D., & Duhnea, C. (2025). When "more" is "too much" – A study on tourists' perception regarding beach restoration in Mamaia on the Romanian Black Sea Coast. *Sustainability, 17*, 922. https://doi.org/10.3390/su17030922

Kim, J. J., Lee, M. J., & Han, H. (2020). Smart hotels and sustainable consumer behavior: Testing the effect of perceived performance, attitude, and technology readiness on word-of-mouth. *International Journal of Environmental Research and Public Health, 17*(20), 7455. https://doi.org/10.3390/ijerph17207455

Liu, X., Wider, W., Fauzi, M. A., Jiang, L., Udang, L. N., & Hossain, S. F. A. (2024). The evolution of smart hotels: A bibliometric review of the past, present, and future trends. *Heliyon, 10*(4), e26472. https://doi.org/10.1016/j.heliyon.2024.e26472

National Institute of Statistics. (2025). TempoOnline database, http://statistici.insse.ro:8077/tempo-online/. Accessed on June 2nd, 2025.

Oklevik, O., Kwiatkowski, G., Malchrowicz-Mośko, E., Ossowska, L., & Janiszewska, D. (2021). Determinants of tourists' length of stay. *PLoS One, 16*(12), e0259709. https://doi.org/10.1371/journal.pone.0259709

Oprea, D. I., & Briciu, A.-E. (2023). Some current data and observations regarding the ski slopes in Suceava County, Romania. *GEOREVIEW, 33*(1), 174–185. https://doi.org/10.4316/GEOREVIEW.2023.01.13

Panyadee, C., Krajangchom, S., Sangkakorn, K., & Intawong, K. (2023). Smart wellness technology for tourism destination based on evolving tourist expectation model. *TEM Journal, 12*(2), 1218–1226. https://doi.org/10.18421/TEM122-68

Polderman, A., Haller, A., Viesi, D., Tabin, X., Sala, S., Giorgi, A., Darmayan, L., Rager, J., Vidovič, J., Daragon, Q., Verchère, Y., Zupan, U., Houbé, N., Heinrich, K., Bender, O., & Bidault, Y. (2020). How Can Ski Resorts Get Smart? *Transdisciplinary Approaches to Sustainable Winter Tourism in the European Alps. Sustainability, 12*(14), 5593. https://doi.org/10.3390/su12145593

Popescu, A., Mărcuţa, A., Mărcuţa, L., Tindeche, C., Hontus, A. & Pleşoianu, D. (2020). Romania's Tourism Trends and Characteristics on the Seaside Resorts of the Black Sea in the Period 2010-2019. *Scientific Papers Series Management, Economic Engineering in Agriculture and Rural Development, 20*(4).

Romanian Government/Guvernul României, Hotărâre de Guvern nr. 852 din 13 august 2008, pentru aprobarea normelor şi criteriilor de atestare a staţiunilor turistice/Governmental Decision (GD) no 852 of August 13th, 2008 regarding the approval of the norms and criteria for tourist resorts' attestation, actualizată prin HG nr. 108 din 16 februarie 2024/updated via GD no 108 of February 16th, 2024. [online] https://legislatie.just.ro/Public/DetaliiDocument/96733. Accessed 12 May 2025.

Romanian Government/Guvernul României, Ministerul Antreprenoriatului şi Turismului (MAT)/Ministry of Entrepreneurship, and Tourism (MET). (2023). *Strategia Naţională a României pentru Dezvoltarea Turismului 2023–2035.* [online] https://turism.gov.ro/web/wp-content/uploads/2022/11/SNRDT-actualizat-var-pt-HG-2023-2035.pdf. Accessed 12 May 2025.

Romanian Government/Guvernul României, Ministerul Economiei, Antreprenoriatului şi Turismului (MEAT)/Ministry of Economy, and Entrepreneurship, and Tourism (MEET). (2025a). *Lista localităţilor atestate ca staţiuni turistice de interes naţional, respectiv local/The List of Localities Attested as Tourist Resorts of National and Local Interest.* [online] https://se.situr.gov.ro/OpenData/OpenDataList?type=listaStatiuni. Last updated on June 2nd, 2025, accessed 2nd June 2025.

Romanian Government/Guvernul României, Ministerul Economiei, Antreprenoriatului şi Turismului (MEAT)/Ministry of Economy, and Entrepreneurship, and Tourism (MEET). (2025b). *Lista pârtiilor de schi omologate/The List of Approved Ski Slopes.* [online] https://se.situr.gov.ro/OpenData/OpenDataList?type=listaPartii. Last updated on June 12th, 2025, accessed 12 June 2025.

Romanian Smart City Association/Asociaţia Română Smart City (ARSC, 2025), *Smart City România. Ghid complet—Domeniile de implementare: verticalele smart city* [online] https://arsc.ro/smart-city-romania/. Accessed 12 May 2025.

Romão, J. (2021). *Tourism, smart specialization and sustainable development.* MDPI. https://doi.org/10.3390/books978-3-03943-672-9

Shen, S., Sotiriadis, M., & Zhang, Y. (2020). The influence of smart technologies on customer journey in tourist attractions within the smart tourism management framework. *Sustainability, 12*, 4157. https://doi.org/10.3390/su12104157

Sigala, M. (2018). New technologies in tourism: From multi-disciplinary to anti-disciplinary advances and trajectories. *Tourism Management Perspectives, 25*, 151–155. https://doi.org/10.1016/j.tmp.2017.12.003

Tache, A. V., Petrişor, A.-I., Ciogescu, O.-E., & Tache, M.-M. (2022). Intermodal GIS analysis model for assessing the travel cost to tourist resorts in Romania. *Cactus – Tourism Journal, 4*(2), 35–45. https://doi.org/10.24818/CTS/4/2022/2.03

Trần Thị Xuân, V., Ha Nam Khanh, G., & Tạ Văn, T. (2024). *Analyzing the impact of smart tourism application, destination image, and satisfaction on tourist intention to return to the tourist destination of Nha Trang*". [online] SSRN: https://ssrn.com/abstract=5000972 or https://doi.org/10.2139/ssrn.5000972.

Volkmann, C., Tokarski, K. O., Dincă, V. M., & Bogdan, A. (2021). The impact of COVID-19 on Romanian tourism. An explorative case study on Prahova County, Romania. *Amfiteatru Economic, 23*(56), 196–205. https://doi.org/10.24818/EA/2021/56/196

World Bank. (2019). *Output 3.2 "Final version of the methodology for developing strategic documents and its application" Romania – National tourism development strategy 2019–2030.* Volume 1: Sector Rapid Assessment Report, February 2019. https://documents.worldbank.org/en/publication/documents-reports/documentdetail/678871650278451023/output-3-2-sector-rapid-assessment-report

Chapter 15
The Importance of Smart Technologies in Tourism Businesses. Toward Metaverse and Tourism Data Space

Tomáš Gajdošík and Beáta Vaneková

Abstract The topic of smart tourism businesses is so far under-researched. Smart tourism businesses are the critical element of the smart tourism ecosystem; however, little is known about technology adoption and data sharing to strengthen their smart approach. The chapter aims to analyze the importance of smart technologies in hospitality and catering facilities, travel agencies, cable car businesses, and sport and recreational facilities for enhancing tourist experience and performance of the business. It also analyzes data collection and sharing in tourism businesses. The chapter comprehensively analyzes the smart tourism concept from the business perspective. The chapter uses quantitative and qualitative research methods. Based on the classification of technologies on stimulating extra-sensory, hyper-personalized, and beyond-automation experiences, the chapter outlines future directions on the use of the metaverse in tourism businesses. Moreover, the research identifies that the participation of businesses in the tourism data space can reduce the barriers to data sharing among businesses.

Keywords Metaverse · Smart businesses · Smart technologies · Tourism data space

15.1 Introduction

The dynamic environment, which is related to technological development, climate change, and growing uncertainty, significantly influences tourism development. This development is conditioned by fragile natural and cultural environments that

---

T. Gajdošík (✉) · B. Vaneková
Faculty of Economics, Matej Bel University in Banská Bystrica, Banská Bystrica, Slovakia
e-mail: tomas.gajdosik@umb.sk

A. L. Negrusa, M. M. Coroş (eds.), *Innovation, Sustainability, and Growth in a VUCA Environment*, Springer Proceedings in Business and Economics,
https://doi.org/10.1007/978-3-032-11639-0_15

must be preserved, as they are the source of its growth. The concept of smart tourism development emerged as a response to changes such as climate change and growing uncertainty. This concept needs to be perceived as a philosophy of tourism development that creates opportunities for (1) better satisfying the needs of visitors with tailor-made experiential offerings, (2) creating new business models leading to the development of smart tourism businesses, and (3) more professional management of tourism destinations (Gretzel et al., 2015).

The business level of smart tourism is underpinned by access to shared data, which facilitates inter-organizational cooperation and the pooling of resources (Xiang & Fesenmaier, 2017a). It encompasses the utilization of both internal datasets and advanced technologies to enhance marketing strategies, profitability, and competitive advantage, as well as the integration of external data sources enabled through data-sharing practices (Buhalis & Leung, 2018). In this sense, it is crucial to provide information on how tourism businesses perceive smart technologies and what the barriers to their implementation are. Moreover, as smartness is not only aimed at technology adoption, it is worth examining the data collection and sharing ethos in tourism businesses. The topic of smart tourism businesses is so far under-researched. Tourism businesses are practically forgotten in the academic literature dealing with smart tourism, as they often lag behind in adopting technological innovations (Ballina, 2022).

15.2 Characteristics of Smart Tourism Businesses

The physical presence of a smart business is closely linked to the virtual environment (Gretzel, 2016). The goal of the smart approach in a business environment is (1) to enhance the visitor's experience, (2) optimize operational processes (increase work productivity and reduce employee burden), and (3) more efficient strategic management (planning and marketing) and business economics (increasing revenue and reducing costs). There are several definitions of smart tourism businesses (see, e.g., Gretzel, 2016; Jeremen et al., 2016; Li et al., 2017; Ballina, 2022; Buhalis et al., 2023b; O'Connor, 2023; Song et al., 2024). The definitions of smart businesses stress the role of smart technologies and data sharing.

Smart technologies refer to general and specific applications designed to enhance tourist experiences, create added value, and improve the competitiveness of tourism destinations and businesses (Neuhofer et al., 2015). The concept of smartness leverages the interconnectivity and interoperability of integrated technologies. These technologies collect and utilize data from diverse sources, including physical infrastructure, social networks, and other sources, and are supported by advanced analytical tools such as machine learning and artificial intelligence. The term *smart tourism technologies* is a broad and somewhat ambiguous umbrella concept that refers to a wide array of advanced technologies aimed at deepening digital connectivity. According to Gajdošík and Orelová (2020), these technologies can be categorized into four main groups: (1) technologies providing information, (2) technologies related to data storage and management, (3) technologies supporting experience, and (4) sensing and network technologies (see Fig. 15.1).

The business dimension of smart tourism development should be grounded in a data-sharing approach that fosters collaboration and resource integration among tourism businesses (Xiang & Fesenmaier, 2017b). This involves the adoption of advanced technologies within tourism businesses to support data collection and operational processes, alongside a strategic commitment to sharing data with other stakeholders. A "smart tourism business" is characterized by its ability to seamlessly integrate internal and external applications, facilitate cloud-based data exchange, combine real-time and historical data, and operate within interconnected and interoperable system environments (Buhalis & Leung, 2018). Tourism businesses employ a range of information systems, including property management systems (PMSs), point of sale systems, sales and management systems, customer relationship management (CRM) systems, and revenue management (RM) systems. Transactional data generated by these systems should be complemented by big data derived from internal sensors, such as motion, voice, and temperature sensors in hotel rooms, or localization sensors in reception areas and restaurants, and external sensors, including those that monitor weather conditions, traffic flow, and social media content. Effective data exchange among business partners and broader stakeholders, such as transportation providers, reservation platforms, event organizers, destination management organizations, and other tourism businesses, is also essential for realizing the full potential of smart tourism systems.

However, so far, the studies on smart tourism have explored mainly the tourist perspective (e.g., Femenia-Serra et al., 2019; Gajdošík, 2020) and the destination level (e.g., Buhalis & Amaranggana, 2014; Gretzel et al., 2015) but not adequately address the business level (e.g., Ballina, 2022; Buhalis & Leung, 2018). To fulfill

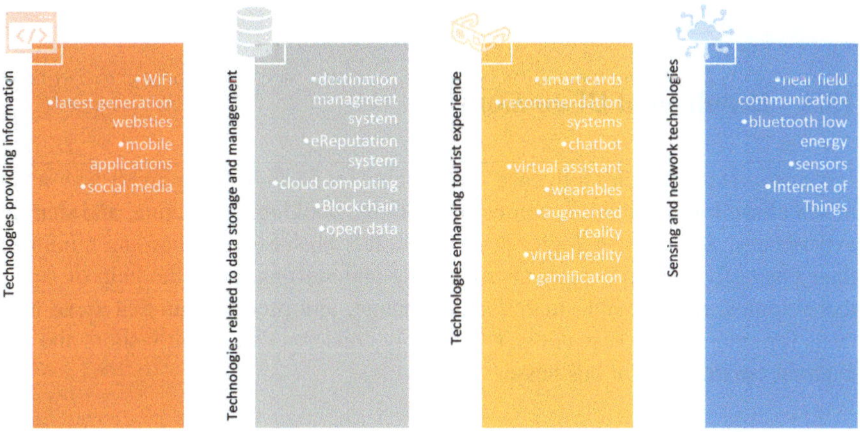

Fig. 15.1 Characteristics of smart technologies. (Source: Own elaboration based on Gajdošík and Orelová (2020))

this gap, the presented study examines the smart tourism businesses from the perspective of smart technologies and data-driven practices.

15.3 Research Methodology

The chapter aims to analyze the importance of smart technologies in hospitality and catering facilities, travel agencies, cable car businesses, cultural facilities, and sport and recreational facilities for enhancing tourist experience and performance of the business, as well as their data-sharing practices. The research sample consists of 310 tourism businesses from the Central European country - Slovakia, whose managers took part in the questionnaire survey. The managers of these tourism businesses were asked to evaluate the importance of selected smart technologies for tourist experience and the effect on business performance on a scale of 1–5 (1—lowest score and 5—highest score). Moreover, they were asked to think of barriers to technology implementation, as well as the data they collect and share about tourists and the external environment.

The survey was conducted online using the computer-assisted web interviewing (CAWI) method in the years 2021 and 2022. The CAWI method represents a dominant paradigm in data gathering in social research (Tomaselli et al., 2022). The e-mail with a link to a web survey was sent to Slovak tourism businesses. The questionnaire was adapted to the specificities of each business category. The research sample consists of 118 independent hotels, 69 restaurants, 33 travel agencies, 31 cable car businesses, 39 cultural institutions, and 20 thermal spas and aquaparks. The analysis is based on the quantitative (descriptive statistics) and qualitative (content analysis) research methods. Slovakia was chosen as a reference country, where technology adoption is at the average level of European Union countries. Therefore, the results present the typical European perspective.

15.4 Results and Discussions

Information technologies influence visitor behavior because they provide new ways of searching for information, comparing options, booking, connecting, and sharing experiences. Smart technologies positively affect business revenues and minimize their costs, for example, in terms of energy and staffing. With the help of smart technologies, it is possible to collect, exchange, and process data that serve as a basis for smart decision-making. This mainly concerns data about visitors and the external environment of businesses.

The Importance of Smart Technologies for Tourism Businesses

As the goal of the smart approach in tourism businesses is to enhance the visitor experience and optimize operational and strategic processes, the analysis is focused on determining how managers perceive the use of smart technologies. Given that different types of tourism businesses require different types of information technologies, it is appropriate to address this issue separately for each type of business. Consequently, the managers should think of the barriers to technology implementation, as well as the data sources they use.

Hotel managers identified that the greatest impact on the visitor experience is made by WiFi, the latest generations of websites, social media, and mobile applications. Also significant are online check-in, CRM system, recommendation system, virtual and augmented reality, chatbot, and online assistance (Fig. 15.2). These technologies provide tourists with current information, facilitate real-time communication with the hotel, and serve as a basis for creating a personalized product. In addition to these technologies, information systems also have a significant impact on hotel performance. From the managers' responses, it can be observed that they attach great importance to information technologies, which positively influence the environmental pillar (systems for measuring energy and waste consumption) and the economic pillar of sustainable development (RM systems, electronic reputation, and latest hotel PMS). A trend in hotel information systems is primarily the use of cloud computing and dynamic pricing. Hotel managers are gradually becoming aware of the importance of Blockchain technology for data protection, as well as wearable devices (social pillar), which guests are increasingly using. Less

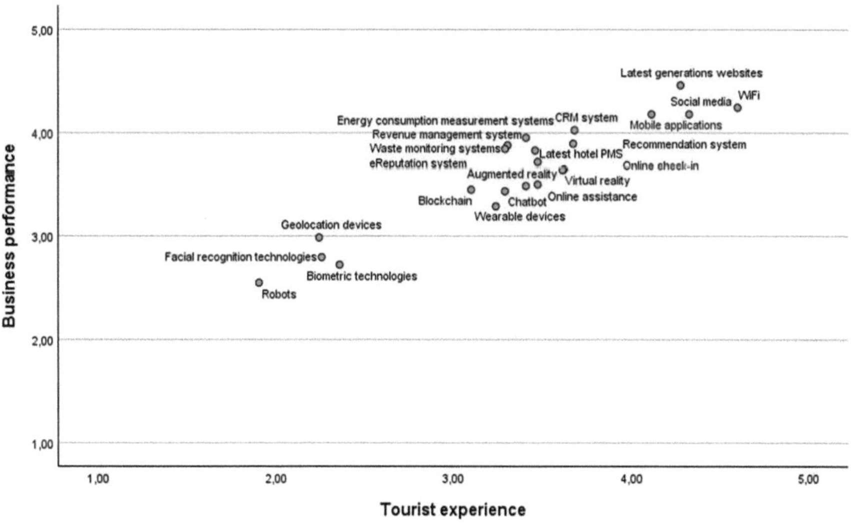

Fig. 15.2 Hotel managers' perspectives on smart technologies. (Source: Authors' own processing based on processing of the collected data)

significant are geolocation devices (e.g., beacons using BLE), which allow monitoring of guest movements within the hotel. By being able to identify the visitor's location in a specific part of the hotel, the visitor can receive personalized alerts (push notifications), enabling them to have a better experience. Technologies for facial recognition, other biometric technologies, and robots are likely to have a greater impact on the hotel industry in the future.

Restaurants utilize fewer technologies than hotels. In these businesses, the highest impact on the tourist experience is mainly from WiFi, social media, mobile applications, and the latest generations of websites, which, like in hotels, provide customers with current information (Fig. 15.3). Especially in fast food establishments, tablets are also significant, speeding up the ordering of meals and drinks. QR codes are also important for guests, as they most often allow them to choose from an electronic menu or rate the work of staff. The performance of these facilities is also influenced by the latest versions of PMSs. The latest versions of these systems should be based on cloud computing, which allows easier integration of additional systems. Together with the use of the Internet of Things, catering facilities can efficiently manage resources. These technologies actively and automatically monitor processes without the need for human intervention. The Internet of Things can monitor storage and inventory status, thereby preventing unnecessary food waste. In production, it can monitor the wear and tear of equipment and automatically report its maintenance. Other technologies are not as significant from the perspective of managers. However, robots are gradually being introduced to help minimize manual labor in production and sales, and augmented reality is being used to enhance the experience of customers waiting for food and drinks and to enhance the consumption experience.

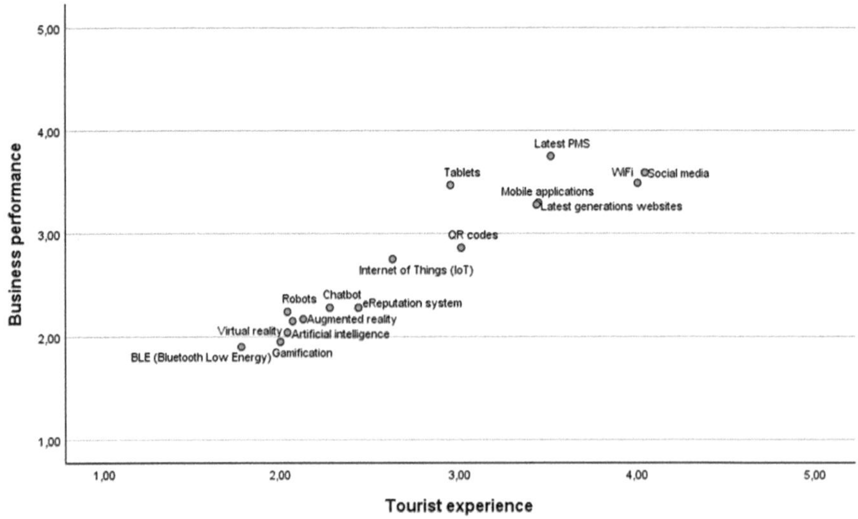

Fig. 15.3 Catering facilities managers' perspectives on smart technologies. (Source: Authors' own processing based on processing of the collected data)

In overall, managers of *travel agencies* rated the significance of smart technologies as high. From the perspective of tourist experience, the most important are the latest generations of websites, the latest PMS, WiFi in travel agency premises, social media, contactless payment, as well as mobile applications and online assistance (Fig. 15.4). These technologies also have the greatest impact on the performance of travel agencies. However, managers are aware of the need to differentiate themselves in the market. Given that tourists increasingly utilize information technologies, managers consider virtual and augmented reality to be significant in terms of both tourist experience and the performance of the travel agency. Simulating a virtual environment allows travel agencies to partially materialize the intangible product of tourism. Through virtual environments, customers can "try out" a tour by simulating images, sounds, smells, and sensations, which can help prevent potential disappointments. The performance of travel agencies is also influenced by the introduction of an electronic reputation system, as well as connections to global distribution systems (GDSs). Robots, chatbots, gamification, and 3D printing in the activities of travel agencies are not currently as significant. Although they have great potential, managers perceive their significance as below average. In the future, the impact of 3D printing can be expected, as it allows for the printing of personalized souvenirs along with a tourist guide at the destination. Such a souvenir can be an authentic alternative to mass-produced souvenirs.

Cable car businesses operate in mountain resorts, which constitute a significant part of tourism offerings in both domestic and international markets. Managers of these businesses consider the latest generation of websites, social media, WiFi, mobile applications, as well as smart cards and NFC technology to be the most important technologies (Fig. 15.5). Smart cards with integrated Near Field

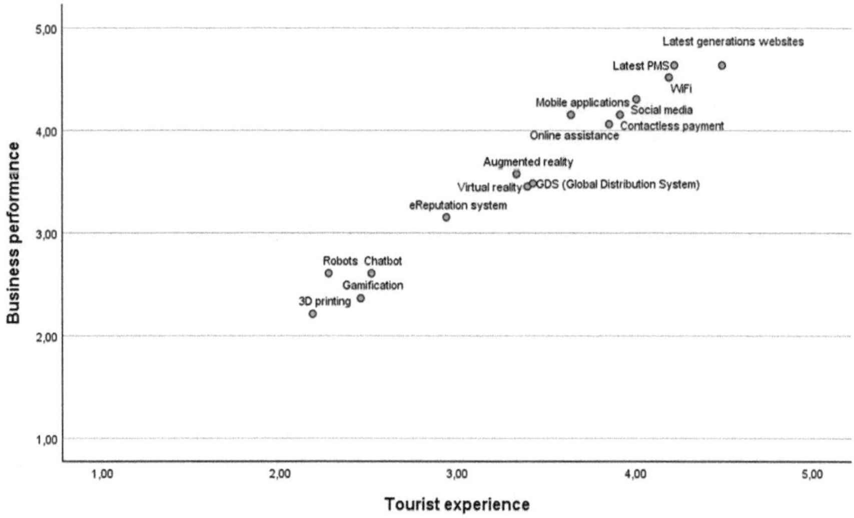

Fig. 15.4 Travel agencies managers' perspectives on smart technologies. (Source: Authors' own processing based on processing of the collected data)

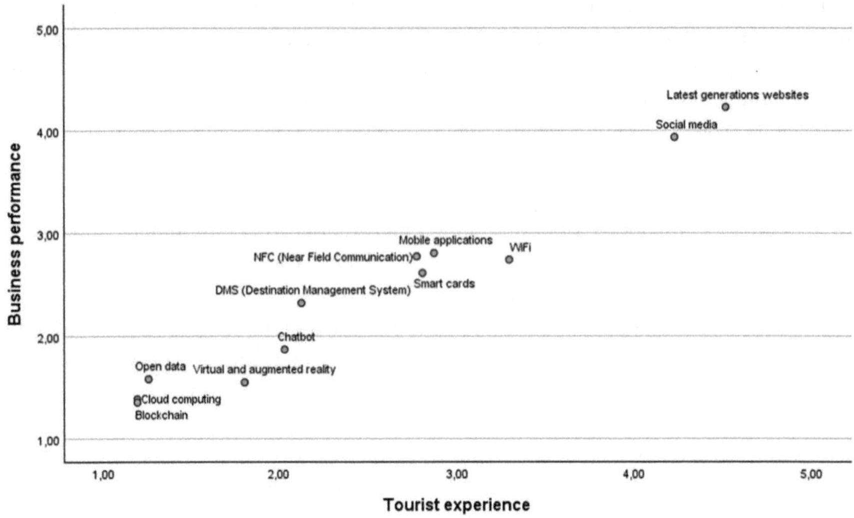

Fig. 15.5 Cable car facilities' managers' perspectives on smart technologies. (Source: Authors' own processing based on processing of the collected data)

Communication (NFC) or Radio-Frequency Identification (RFID) enable contactless payments or terminal access. They are also a good source of visitor data, as they can monitor consumption, visited attractions, and the movement of visitors around the destination. It is advantageous if cable car businesses are integrated within the destination management system. This system allows connecting stakeholders in the destination and informing them about sustainable development policies while also serving as a source of current information for visitors. Chatbots, virtual and augmented reality, open data, cloud computing, and Blockchain have less significance. An example is the use of a smart skiing helmet with integrated augmented reality technology. The helmet camera captures the surroundings on the slope and provides skiers with visual information such as altitude or speed. Additionally, the helmet is equipped with headphones and a microphone, allowing communication with friends via text, sound, or video. In the future, these benefits could be expanded to include information such as lift wait times and slope congestion.

Cultural facilities provide their services not only to residents but also to tourists. Based on the responses of managers of these facilities, information technologies have an impact on enhancing tourist experience as well as the performance of these organizations. In addition to traditional technologies significant in other types of tourism businesses, cultural facilities place great importance on QR codes, holograms, and gamification (Fig. 15.6). QR codes allow visitors of cultural facilities (such as museums or galleries) to access more information about exhibited objects by redirecting them to the facility's website, where this information is available. This way, the exhibition space is not visually cluttered with information, and visitors who wish to learn more have a simple means to obtain the desired information. Information technologies also enable addressing problems of cultural facilities

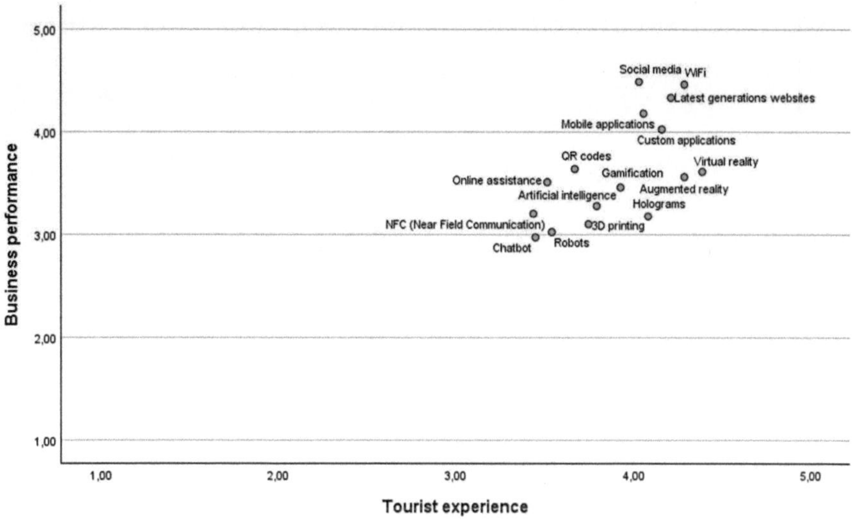

Fig. 15.6 Cultural facilities managers' perspectives on smart technologies. (Source: Authors' own processing based on processing of the collected data)

when their objects are not accessible to visitors due to space constraints or a high likelihood of their deterioration. In such cases, it is appropriate to use holograms, which create a three-dimensional image of the object based on projection. By utilizing gamification, cultural facilities have the opportunity to increase visitor engagement and build a loyal relationship with them.

The core infrastructure conditions for active leisure and recreational sports are *sports and recreational facilities*. Among the most utilized sports and recreational facilities are thermal pools and water parks. According to their managers, WiFi availability within the facility is of highest importance. Additionally, mobile applications, the latest generations of websites, electronic reputation systems, and QR codes are significant (Fig. 15.7). Managers also emphasize the importance of utilizing RFID technology, which is integrated into visitor wristbands. Such wristbands enhance safety and convenience, allowing visitors to purchase food, beverages, or souvenirs without concerns about losing money or having it stolen. Currently, less emphasis is placed on big data processing, chatbots, cloud computing, augmented and virtual reality, as well as gamification. However, these technologies have great potential to impact tourist experience and facility performance. Incorporating gaming elements into water slides, for instance, motivates visitors to revisit the facility. These gaming elements offer visitors new experiences, competition, and excitement. In some water parks, visitors can touch LED light panels embedded in the inner side of the slide, earning points, thus fostering competitiveness and new experiences. Virtual reality is also utilized during water slide rides, for example, in the Prague aquapark, where visitors can wear waterproof gear and experience the slide in a completely different environment. Similarly, augmented reality allows for the

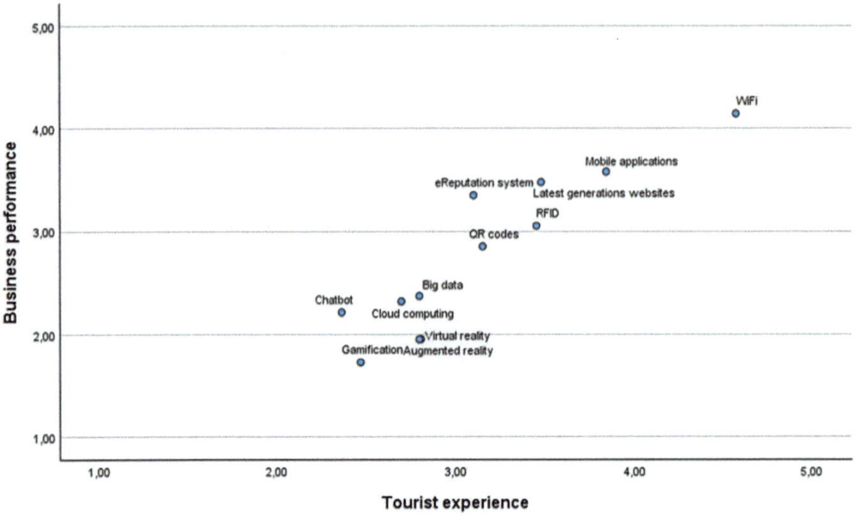

Fig. 15.7 Sports and recreational facilities managers' perspectives on smart technologies. (Source: Authors' own processing based on processing of the collected data)

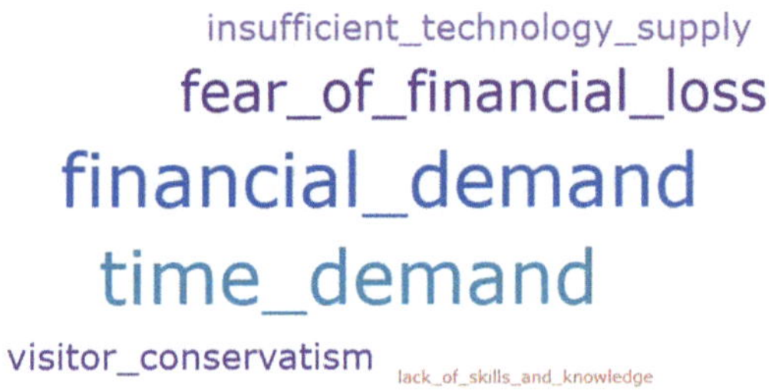

Fig. 15.8 Barriers to the implementation of smart technologies in the tourism business. (Source: Authors' own processing based on processing of the collected data)

display of real rides enriched with interactive graphics. Visitors can thus see their current speed or compete simultaneously with multiple riders on the slide.

Managers of the analyzed businesses identified the financial demands of implementing such technologies as the most significant barrier to adopting smart technologies (Fig. 15.8). The initial costs for smart technologies range in the tens of thousands of euros. For example, based on the answers of restaurant managers, purchasing a robot would cost a restaurant approximately 17,000 euros. In addition to these initial financial costs, maintenance and potential repairs need to be considered. Therefore, it is often suitable to address such initial investments through leasing, where the monthly installment, including maintenance, is approximately 5000 euros. This financial challenge is particularly critical for small- and medium-sized tourism businesses, which often lack the capacity to invest in advanced technologies independently. To help lower this entry barrier, support in the form of shared digital platforms, training initiatives, or regional innovation hubs can be essential. Additionally, partnerships with universities or technology providers can offer low-cost opportunities for experimentation and knowledge transfer, making innovation more accessible even for resource-constrained businesses.

The second most frequently mentioned barrier is fear of losing financial resources, associated with risk aversion. Managers are concerned that the technology will not yield the desired benefits, resulting in the business unnecessarily expending financial resources. This risk can be mitigated by adjusting contractual terms or conducting a thorough analysis of case studies where these technologies have been implemented.

According to managers, another barrier to the implementation of smart information technologies is the conservatism of visitors, who find the introduction of technologies in providing tourism services obstructive. Therefore, in several businesses, these technologies have begun to be introduced as "calm," with a design that respects the customer's perspective (Stankov et al., 2019). Within the calm design, technologies work in the background, and their operation is only visible to the customer when necessary. Due to these technologies, employees perform fewer repetitive tasks, businesses have better customer information, and they have more time to focus on customer needs. Thanks to such technologies, it is possible to continuously increase personal contact, which is important in tourism. Managers of the analyzed businesses also identified time-consumption and insufficient technology offerings as barriers. These barriers are gradually minimized through the progressive development and improvement of technologies.

Furthermore, the analyzed importance of technologies was classified based on Buhalis et al. (2019) into technologies supporting extrasensory experiences (e.g., virtual and augmented reality), technologies supporting hyper-personalized experiences (e.g., sensors, Internet of Things, and localization devices), and technologies supporting beyond-automatic experiences (e.g., chatbots, virtual assistants, and robots). Based on the identified average importance of technologies by tourism businesses, technologies supporting extrasensory experiences have the greatest significance.

For hotels, travel agencies, and cultural facilities, technologies supporting extrasensory experiences are of highest importance. Catering facilities, cable car businesses, and sports and recreational facilities prioritize hyper personalized experiences (Table 15.1).

Based on the significance of disruptive technologies for tourism businesses, efforts to develop tourism in the metaverse can be supported. As the metaverse represents a predominantly extra-sensory paradigm, its applications are particularly relevant for hotels, cultural institutions, and tour operators, sectors in which the enhancement of sensory experiences is considered especially critical. The metaverse influences the entire spectrum of tourist behavior, spanning the pre-travel planning stage, the in-destination experience, and post-travel engagement. Within the context of smart tourism, the integration of metaverse technologies signifies a paradigm shift, enabling the creation of immersive and enriched experiences, the convergence of physical and virtual environments, and the development of innovative modes of interaction between tourists and tourism businesses. It enables tourism businesses to collaborate with other stakeholders across the entire ecosystem,

Table 15.1 The importance of disruptive technologies by tourism businesses

Type of tourism business	Extrasensory experiences	Hyper-personalized experiences	Beyond-automatic experiences
Hotels	3.56	2.48	2.79
Catering facilities	2.10	2.27	2.15
Travel agencies	3.44	2.20	2.50
Cable car businesses	1.68	2.71	1.95
Cultural facilities	3.96	3.37	3.25
Sport and recreational facilities	2.38	2.58	2.29
TOURISM BUSINESSES (mean)	3.00	2.54	2.56

Source: Authors' own processing based on processing of the collected data

leading to an intensified visitor experience (Buhalis et al., 2023a). At the same time, tourism businesses and service providers should leverage the metaverse to create personalized virtual and physical experiences that take into account multiple avatars or user profiles (Koo et al., 2023). By encouraging potential visitors to try out products and services in the online environment of the metaverse, tourism service providers can showcase the unique features of their facilities and destinations, thereby inspiring interest in trying these products, services, and ultimately visiting the destinations in real life (Chen & Yao, 2022; Tsai, 2024). However, the metaverse technologies and their use in tourism are still in their infancy. Besides the barriers to the implementation of smart technologies, there are specific barriers to metaverse implementation. They are usage barriers (inconvenience and unclear content), value barriers (high cost of technology), risk barriers (lack of trust), traditional barriers (unpleasant experience), image barriers (difficulty to use), technology vulnerability barriers (technology anxiety), and ideological barriers (skepticism) (Sowmya et al., 2024). From a technological perspective, these barriers can be reduced by improving graphics, implementing better sensors (including touch, taste, smell, and temperature), and introducing better interactivity, more intuitive interfaces, and easier, user-friendly navigation within the virtual environment.

Data Collection and Sharing by Businesses

Smart technologies enable data collection and processing. Data on tourists are crucial for tourism businesses. Up to 77% of the analyzed businesses collect basic data about tourists (name, age, gender, and email). These data are used for demographic segmentation and contacting tourists to gather feedback and motivate them to reuse services. More than half of the analyzed businesses collect data on tourist spending as a basis for RM. Only 26% of businesses are interested in preferences (hobbies and interests), which serve as the basis for creating personalized products.

One-fourth of the businesses also collect other data, most commonly website data (Fig. 15.9). In hotels, the other data includes guest movement within the hotel and booking methods. Travel agencies collect data on the popularity of destinations. Cable car businesses collect data on the travelled distance, while thermal pools and water parks collect data on attraction popularity. Such data enable tourism businesses to predict demand for their services, optimize offerings (e.g., website content), and prepare tailored products. They also allow for injecting a dynamic element into business marketing, positively influencing their competitiveness.

Data about the external environment is also crucial. Business managers should monitor the environment in which their business operates in order to proactively adapt their decisions to changing circumstances. The analyzed businesses most commonly monitor the pricing policies of their competitors (Fig. 15.10). They also collect data on the organization of major events in their vicinity. These events positively influence the demand for business services. Up to 59% of the analyzed businesses monitor the key performance indicators of competing businesses. These typically include occupancy rates, reservations, and revenues. Sources of information include online travel agencies and GDSs. Aggregated data on accommodation and transportation reservations, event organization within the destination, and holidays and festivals in the vicinity are important data inputs for RM as they provide information on anticipated demand. To a lesser extent, tourism businesses gather data on market trends, the security situation at the destination, long-term weather forecasts, and transportation price developments. This data is stored in their internal systems. However, the smart hospitality framework requires full integration of internal and external applications, cloud-based data exchange, and access to both real-time and historical data from internal and external sources (Buhalis & Leung, 2018; Buhalis et al., 2023b).

When implementing the smart approach in tourism businesses, the exchange of data, information, and knowledge among these businesses is also crucial. Therefore, the analyzed businesses were asked to provide the main barriers to data sharing. The most significant barriers to data sharing in tourism businesses are the lack of trust, fear of data misuse, and reluctance to communicate. These barriers can be overcome by higher participation of businesses in tourism data spaces. The most important role is played by destination management organizations and event organizers. Tourism businesses generate knowledge, but they are not considered the most important stakeholders. These findings support the research of Raisi et al. (2020), who state that small- and medium-sized businesses are not central actors in knowledge sharing and rely on a limited number of hubs. For this reason, it is necessary to stimulate their cooperation and trust in order to expand the flow of knowledge and promote the smart tourism development.

The establishment of a data space within the tourism sector has the potential to foster a culture of data sharing among stakeholders (DATES, 2023). These dynamic data ecosystems enable the exchange and flow of data between diverse actors, including public authorities, tourism businesses, and tourists themselves (European Commission, 2023). As part of its broader European Data Strategy, the European Commission seeks to develop a common European tourism data space, with the

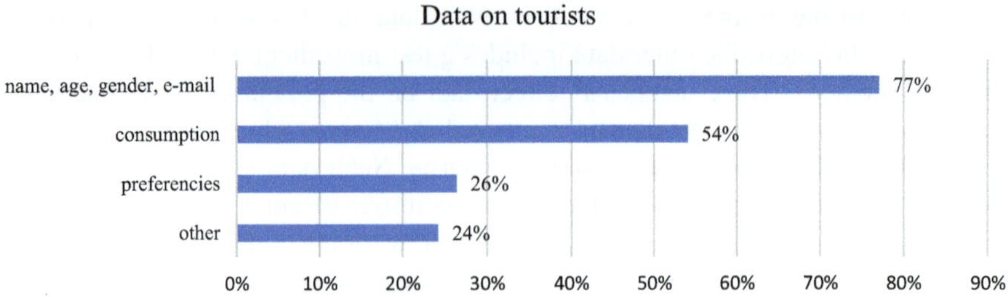

Fig. 15.9 The most important data on tourists in tourism businesses. (Source: Authors' own processing based on processing of the collected data)

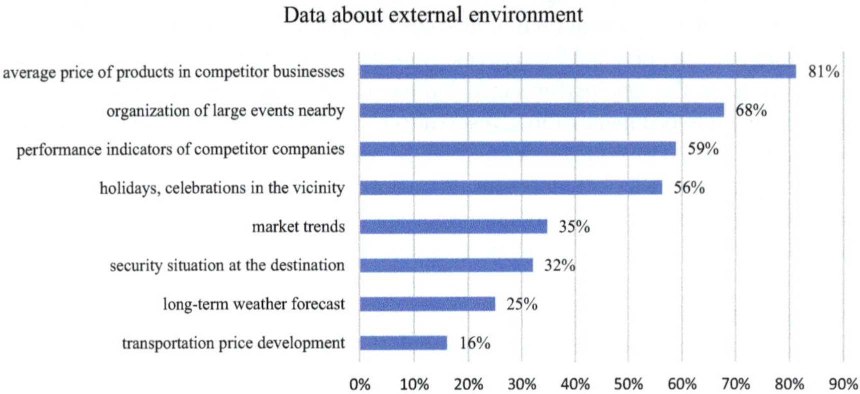

Fig. 15.10 The most important data about the external environment in tourism businesses. (Source: Authors' own processing based on the processing of the collected data)

objective of building trust among participants, encouraging collaboration and innovation, and supporting the creation of data-driven tourism policies, services, and products (Stienmetz & Kolomoyets, 2024). A critical challenge in the implementation of such a data space concerns the effective sharing of information, underscoring the importance of establishing clear governance frameworks, defining technical standards, and ensuring equitable participation and benefit-sharing among the involved stakeholders (Ordóñez-Martínez et al., 2024). The willingness to participate in tourism data space can be improved by peer learning activities, outsourcing of data analytical skills, and establishing data monetization schemes and interoperability standards (Gajdošík & Marciš, 2025).

15.5 Conclusions

The research discusses the under-researched area of smart tourism businesses. These businesses are often overlooked in discussions about smart tourism due to their slower adoption of technological innovations (see Ballina, 2022). Different types of tourism businesses prioritize different aspects of smart technologies, with some focusing on extrasensory experiences and others on hyper-personalized ones. In this context, the metaverse presents a significant opportunity for enriching tourist experiences and bridging physical and virtual realms, particularly for hotels, cultural facilities, and tour operators. Some tourists may be resistant to the introduction of smart technologies, requiring businesses to implement them in a calm manner that respects customers' perspectives (Stankov et al., 2019). Moreover, the metaverse is still in its infancy, and several technical advancements are needed (e.g., improving graphics, better sensors, and interactivity) to fully embrace its potential.

Furthermore, data collection, including basic demographic data and preferences, is crucial for tourism businesses. However, there is room for improvement in collecting and utilizing more detailed data for personalization. Tourism businesses face several barriers in data sharing, mainly a lack of trust, fear of data misuse, and reluctance to communicate. These barriers can be overcome by promoting the participation of businesses in the tourism data space.

The presented findings point toward the metaverse and the emerging concept of the tourism data space as key technological frameworks with significant potential to stimulate the smart tourism businesses (Fig. 15.11). The metaverse, through improved graphics, better sensors, and interactivity, offers new avenues for enhancing tourist engagement, co-creating experiences, and bridging physical and virtual

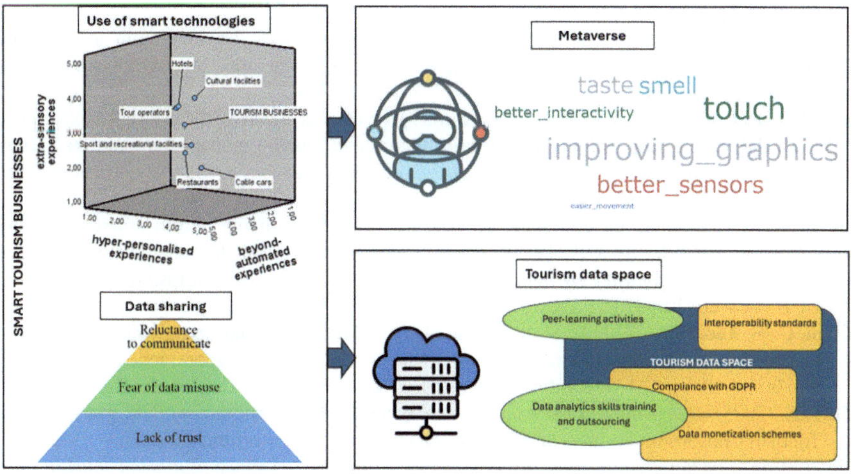

Fig. 15.11 Metaverse and tourism data space fostering smart tourism businesses. (Source: Authors' own processing)

tourism services. Simultaneously, the development of a tourism data space fosters a structured and interoperable environment for data sharing among stakeholders. Integrating peer-learning activities and training in data analytic skills facilitates evidence-based decision-making. Together, these frameworks represent a shift toward more connected, data-driven, and experience-oriented tourism business models.

Future research should explore strategies to foster higher collaboration, trust-building, and secure data practices within the tourism sector. Investigating how participation in tourism data spaces can reduce these concerns and enhance smart tourism development is essential. Additionally, further studies should analyze the practical integration of emerging technologies like the metaverse in different types of tourism businesses, as well as the long-term effects of smart technology adoption on tourist satisfaction and business performance.

Acknowledgments The research was supported by the research project VEGA 1/0136/23 "From resilience to sustainability. The impact of data on sustainable and competitive development of tourism."

References

Ballina, F. J. (2022). Smart business: The element of delay in the future of smart tourism. *Journal of Tourism Futures, 8*(1), 37–54. https://doi.org/10.1108/JTF-02-2020-0018

Buhalis, D., & Amaranggana, A. (2014). Smart tourism destinations. In Z. Xiang & L. Tussyadiah (Eds.), *Information and communication technologies in tourism 2014* (pp. 553–564). Springer International Publishing. https://doi.org/10.1007/978-3-319-03973-2

Buhalis, D., & Leung, R. (2018). Smart hospitality – interconnectivity and interoperability towards an ecosystem. *International Journal of Hospitality Management, 71*(April), 41–50. https://doi.org/10.1016/j.ijhm.2017.11.011

Buhalis, D., et al. (2019). Technological disruptions in services: Lessons from tourism and hospitality. *Journal of Service Management, 30*(4), 484–506. https://doi.org/10.1108/JOSM-12-2018-0398

Buhalis, D., Lin, M. S., & Leung, D. (2023a). Metaverse as a driver for customer experience and value co-creation: Implications for hospitality and tourism management and marketing. *International Journal of Contemporary Hospitality Management, 35*(2), 701–716. https://doi.org/10.1108/IJCHM-05-2022-0631

Buhalis, D., O'Connor, P., & Leung, R. (2023b). Smart hospitality: From smart cities and smart tourism towards agile business ecosystems in networked destinations. *International Journal of Contemporary Hospitality Management, 35*(1), 369–393. https://doi.org/10.1108/IJCHM-04-2022-0497

Chen, C., & Yao, M. Z. (2022). Strategic use of immersive media and narrative message in virtual marketing: Understanding the roles of telepresence and transportation. *Psychology & Marketing, 39*(3), 524–542. https://doi.org/10.1002/MAR.21630

DATES. (2023). *Blueprint and roadmap for deploying the European tourism data space* DATES Tourism Consortium Partners and Data Space for Tourism Consortium Partners.

EC. (2023). *Communication from the commision—Towards a common European tourism data space: Boosting data sharing and innovation across the tourism ecosystem* Brussels. 2023.

Femenia-Serra, F., Neuhofer, B., & Ivars-Baidal, J. A. (2019). Towards a conceptualisation of smart tourists and their role within the smart destination scenario. *Service Industries Journal, 39*(2), 109–133. https://doi.org/10.1080/02642069.2018.1508458

Gajdošík, T. (2020). Smart tourists as a profiling market segment: Implications for DMOs. *Tourism Economics, 26*(6), 1042–1062. https://doi.org/10.1177/1354816619844368

Gajdošík, T., & Marciš, M. (2025). Data-driven practices in European smart tourism destinations: Towards a tourism data space. In L. Nixon, A. Tuomi, & P. O'Connor (Eds.), *Information and communication technologies in tourism 2025. ENTER 2025. Springer proceedings in business and economics*. Springer. https://doi.org/10.1007/978-3-031-83705-0_4

Gajdošík, T., & Orelová, A. (2020). Smart technologies for smart tourism development. In R. Silhavy (Ed.), *Advances in intelligent systems and computing* (pp. 333–343). Springer. https://doi.org/10.1007/978-3-030-51971-1_27

Gretzel, U. (2016). The new technologies tsunami in the hotel industry. In M. Ivanova, S. Ivanov, & V. P. Magnini (Eds.), *The Routledge handbook of hotel chain management* (pp. 516–523). Routledge. https://doi.org/10.4324/9781315752532-54

Gretzel, U., et al. (2015). Smart tourism: Foundations and developments. *Electronic Markets, 25*(3), 179–188. https://doi.org/10.1007/s12525-015-0196-8

Jeremen, D. E., Jędrasiak, M., & Rapacz, A. (2016). The concept of smart hotels as an innovation on the hospitality industry market – Case study of Puro Hotel in Wrocław. *Economic problems of tourism, 36*(4), 65–75. https://doi.org/10.18276/EPT.2016.4.36-06

Koo, C., et al. (2023). Metaverse tourism: Conceptual framework and research propositions. *Current Issues in Tourism.*. Routledge, 3268–3274. https://doi.org/10.1080/13683500.2022.2122781

Li, Y., et al. (2017). The concept of smart tourism in the context of tourism information services. *Tourism Management, 58*, 293–300. https://doi.org/10.1016/J.TOURMAN.2016.03.014

Neuhofer, B., Buhalis, D., & Ladkin, A. (2015). Smart technologies for personalized experiences: A case study in the hospitality domain. *Electronic Markets, 25*(3), 243–254. https://doi.org/10.1007/s12525-015-0182-1

O'Connor, P. (2023). Small- and medium-sized tourism enterprises and smart tourism: Tourism Agenda 2030 perspective article. *Tourism Review, 78*(2), 339–343. https://doi.org/10.1108/TR-09-2022-0431

Ordóñez-Martínez, D., Seguí-Pons, J. M., & Ruiz-Pérez, M. (2024). Conceptual framework and prospective analysis of EU tourism data spaces. *Sustainability (Switzerland), 16*(1), 371. https://doi.org/10.3390/su16010371

Raisi, H., et al. (2020). A network perspective of knowledge transfer in tourism. *Annals of Tourism Research, 80*. https://doi.org/10.1016/j.annals.2019.102817

Song, B., et al. (2024). Discovery of smart hotels' competitiveness based on online reviews. *International Journal of Hospitality Management, 123*. https://doi.org/10.1016/j.ijhm.2024.103926

Sowmya, G., et al. (2024). Exploring innovation resistance in tourism: Barriers to metaverse adoption among tourists. *Cogent Business & Management, 11*(1). https://doi.org/10.1080/23311975.2024.2400309

Stankov, U., Filimonau, V., & Slivar, I. (2019). Calm ICT design in hotels: A critical review of applications and implications. *International Journal of Hospitality Management, 82*, 298–307. https://doi.org/10.1016/j.ijhm.2018.10.012

Stienmetz, J., & Kolomoyets, Y. (2024). *Stakeholder requirements and governance for the European data space for tourism* (pp. 278–283). https://doi.org/10.1007/978-3-031-58839-6_30

Tomaselli, V., et al. (2022). Methods, developments, and technological innovations for population surveys. *Social Science Computer Review*. SAGE Publications, 994–1013. https://doi.org/10.1177/0894439321994218

Tsai, S. P. (2024). Investigating metaverse marketing for travel and tourism. *Journal of Vacation Marketing, 30*(3), 479–488. https://doi.org/10.1177/13567667221145715

Xiang, Z., & Fesenmaier, D. (2017a). Analytics in tourism design. In Z. Xiang & D. Fesenmaier (Eds.), *Analytics in smart tourism design, tourism on the verge* (pp. 1–10). Springer International Publishing Switzerland. https://doi.org/10.1007/978-3-319-44263-1_1
Xiang, Z., & Fesenmaier, D. R. (2017b). Big data analytics, tourism design and smart tourism. In D. R. Xiang, Zheng, & Fesenmaier (Eds.), *Annalytics in smart tourism design, concepts and methods* (pp. 299–307). Springer International Publishing Switzerland. https://doi.org/10.1007/978-3-319-44263-1

MIX
Papier aus verantwortungsvollen Quellen
Paper from responsible sources
FSC® C105338

If you have any concerns about our products,
you can contact us on
ProductSafety@springernature.com

In case Publisher is established outside the EU,
the EU authorized representative is:
Springer Nature Customer Service Center GmbH
Europaplatz 3, 69115 Heidelberg, Germany

Printed by Libri Plureos GmbH
in Hamburg, Germany